TECHNOLOGICAL TRANSFORMATION
Contextual and Conceptual Implications

PHILOSOPHY AND TECHNOLOGY
VOLUME 5

Series Editor: PAUL T. DURBIN

Editorial Board

OFFICIAL PUBLICATION OF THE SOCIETY
FOR PHILOSOPHY AND TECHNOLOGY

Technological Transformation

Contextual and Conceptual Implications

Edited by
EDMUND F. BYRNE
Department of Philosophy, Indiana University, Indianapolis

and

JOSEPH C. PITT
Department of Philosophy, Virginia Polytechnic & State University, Blacksburg

Kluwer Academic Publishers
DORDRECHT / BOSTON / LONDON

Library of Congress Cataloging in Publication Data
　Technological transformation : contextual and conceptual implications / edited by
　Edmund F. Byrne, Joseph C. Pitt.
　　　p. cm. — (Philosophy and technology ; v. 5)
　　Papers prepared for a conference on technology transfer and the
　Third World: issues in the history and philosophy of technology,
　Virginia Polytechnic Institute and State University, Blacksburg,
　Virginia, July 15–18, 1987.
　　Includes index.
　　ISBN-13: 978-90-277-2827-2
　　1. Technology—Philosophy—Congresses.　2. Techonology transfer-
　-Congresses.　I. Byrne, Edmund F., 1933- 　.　II. Pitt, Joseph C.
　III. Series.
　T14.T3836 1989　　　　　　　　　　　　　　　　　　　　88–8263
　601—dc19　　　　　　　　　　　　　　　　　　　　　　　　　CIP
ISBN-13: 978-90-277-2827-2　　　　e-ISBN-13: 978-94-009-2597-7
DOI : 10.1007/978-94-009-2597-7

Published by Kluwer Academic Publishers,
P.O. Box 17, 3300 AA Dordrecht, The Netherlands.

Kluwer Academic Publishers incorporates
the publishing programmes of
D. Reidel, Martinus Nijhoff, Dr. W. Junk and MTP Press.

Sold and distributed in the U.S.A. and Canada
by Kluwer Academic Publishers,
101 Philip Drive, Norwell, MA 02061, U.S.A.

In all other countries, sold and distributed
by Kluwer Academic Publishers Group.
P.O. Box 322, 3300 AH Dordrecht, The Netherlands.

printed on acid free paper

TABLE OF CONTENTS

PART II. CROSS-CULTURAL TRANSFORMATION

INTRODUCTION

The philosophical study of technology has acquired only recently a voice in academic conversation. This situation is due, in part, to the fact that technology obviously impacts on "the real world," whereas the favored stereotype of philosophy allegedly does not. Furthermore, in some circles it was assumed that philosophy *ought not* impinge on the world. This bias continues today in the form of a general dismissal of the growing area now referred to as "applied philosophy". By contrast, the academic scrutiny of *science* has for the most part been accepted as legitimate for some 30 years, primarily because it has been conducted in a somewhat ethereal manner. This is, in part, because it was believed that, science being pure, one could think (even philosophically) about science without jeopardizing one's intellectual purity. Since World War II, however, practitioners of the metascientific arts have come to acknowledge that science also shows signs of having touched down on numerous occasions in what can only be identified as the real world. No longer able to keep this banal truth a secret, purists have sought to defuse its import by stressing the difference between pure and applied science; and, lest science be tainted by contact with the world through its applications, they have devoted additional energy to separating applied science somehow from technology.

This was the intellectual milieu when, in the 1970s, some philosophically oriented scholars began to explore the many facets of technology that had once been – and to a great extent still remain – largely the domain of the social sciences (here understood as including history). Being themselves, however, inheritors of the traditions with which they were eventually in various ways to part company, the positions they initially adopted showed the influence of those traditions. Some favored more theoretical, even metaphysical, approaches to the study of technology. Others were more open to "applying" philosophical methodologies, e.g., argument analysis, to problems engendered by the unquestionably complex presence of technology in the world.

These (so perceived) alternative perspectives have in the past several decades experienced considerable cross-fertilization in relation not only

vii

Edmund F. Byrne and Joseph C. Pitt (eds), Technological Transformation: Contextual and Conceptual Implications, vii–xi.
© *1989 Kluwer Academic Publishers.*

to one another but to "mainstream philosophy" as well. What can come of such cross-fertilization is illustrated in a number of very different but comparably fascinating ways in Part One of this volume, about which more later. In addition, however, the dynamics of philosophical attention to technology-in-the-world could not help but lead eventually to acknowledging a heretofore only tacitly acknowledged fact, to wit, that the world in which technology is to be found includes those places – the so-called Third World – where a majority of the human race still lives with considerably less than full participation in the advances, and pitfalls, of mature development.

In the First World, philosophical reflection has not been oblivious to the world beyond what is familiar. But it has been characterized by a fairly common *assumption* to the effect that *people in developing countries must have more or less the same sorts of concerns about technology as do people in developed countries, the differences being more a function of technical prowess than of cultural preference.* This assumption, in fact, could even be read into the title of the conference in connection with which the papers in this volume were produced: Technology Transfer and the Third World: Issues in the History and Philosophy of Technology (Virginia Polytechnic Institute and State University, Blacksburg, Virginia, July 15–18, 1987). "Technology transfer" has historically been the catch phrase for everything that is good and proper about technological progress, including (not just incidentally) the sophisticated weapons that profit a few at the cost of elevating "brush wars" into potentially global conflagrations. For this and other reasons, certain philosophers in developed countries have done their profession no service by suggesting that all the important philosophical questions are being asked in the northern hemisphere. (A particularly notorious example of this mentality is Richard Rorty's comments at the XIth Interamerican Congress). In addition to exhibiting a rather effete form of cultural imperialism, the implication that no important philosophical questions arise out of common day-to-day experience reduces the methodological injunctions that accompany this attitude to mere hyperbole. At the very least, with respect to the continuing source of pressing philosophical problems, philosophers of technology have come to know better, witness their attention to the concerns and aspirations of the appropriate technology (AT) movement.

With such considerations in mind regarding the role of philosophical inquiry, we planned and organized the fourth biennial conference of the

Society for Philosophy and Technology; sought in vain for funding that would enable Third World scholars to attend; then, this lacuna notwithstanding, hoped that somehow their views would be given fair representation. The results are mixed in this respect, but still gratifying as a first step, for three reasons: four contributors, although now working in North America, are from Third World countries; a number of contributors from developed countries have at least spent some time in the developing world; and even those contributors whose life experience has been for the most part in the developed world do call into question various First World biases that tacitly equate a country's capacity for justice with its technical know-how.

The common thread that runs through all these papers we call *technological transformation*. This is a deliberately ambiguous umbrella concept that includes both transformation of and transformation by technology. The process of rethinking what is supposedly "obvious" about these transformations as they impact on the real world brings to the fore what we call their contextual and conceptual implications, which are many and varied.

The papers in Part One progress from consideration of transformations of self and knowledge to transformations in ways of viewing social implications of technology. Along the way, the process of technological development is passed through a prism the different facets of which display different aspects of valuation, value clarification, and justification.

The first three papers suggest that attention to technology transforms our conceptualization of self, knowledge, and epistemology: for Joseph Margolis, the very notion of a knowing self needs to be reconsidered in light of the reality of technology; Tyrone Lai would reinterpret the epistemology of knowledge to take into account what happens when a particular technology of discovery is closely examined; and Paul Durbin contends that even the accepted paradigms of science and engineering need to be reinterpreted when studied in the social context of an R & D setting.

The next four papers seek to transform our assumptions about technology assessment: Philip Shepard thinks the presumed incompatibility between a factual and a normative approach to assessing technology might be overcome by reinterpreting the resulting disagreement through a language-analytic process of "impartial intervention"; J.A. Crane says the fact-value dichotomy (allegedly laid to rest in the 1930s) still survives if only tacitly in RCBA methods of quantification; Alfred Nordmann recommends the "intrinsic" safety standards of the elevator as being more reliable than the more commonly followed "extrinsic" standards of a governor; and James Klagge, turning to the level of ordinary experience, describes a sense of technology assessment which can be characterized as a function of the age of those performing the assessment.

In each of the last three papers in Part One concerns about technology generate questions about political theory, in particular "standard brand" liberalism that disregards the mixed results of new technology: Albert Borgmann argues that Michael Sandel's widely discussed critique of liberalism calls into question its assumption that technology is an unproblematic provider of good things; Clifford Christians argues for a way to evaluate technology that takes into account the non-technical values of people, including people in other countries who may be affected differentially by so-called global technologies; and Edmund Byrne appeals to the interests of local communities around the world to challenge the justice of work restructuring deemed advantageous to a globalizing corporation.

Contributors in Part Two focus on technological transformations in the Third World, and call our attention to a number of special problems that must be taken into account if philosophy of technology is to become more than merely a reflection of what interests and concerns people in developed countries. They agree that factors other than the interests of developed countries are essential to obtaining a good fit for technologies adopted in a Third World country.

The first four papers assume, with provisos, that technological transformation, if well managed, can be a blessing. Stanley Carpenter warns, however, that technology transfer involves not only economic factors but also such broader cultural factors as the political, the religious, the sociological, and in particular, as suggested by the work of Joseph Margolis, how technology functions as a language. Bernard denOuden examines a successful case of appropriate technology in action, in Egypt; and Romualdas Sviedrys considers more complex exchanges, notably of aircraft technology involving Israel. Lan Xue offers an account of technology policy in the People's Republic of China that is remarkable for its familiarity not only with planning issues faced by the Chinese government but with a wide range of Western literature about appropriate technology and related topics.

The last five papers focus more on the negative aspects of technology transfer and, accordingly, call for caution and controls of various sorts. S. Muthuchidambaram says that the structural imbalances of neocolonialism minimize benefits to Third World host countries of technologies introduced there for export of products to the developed world. Friedrich Rapp draws upon the history of technology in the West to urge that the cultural impact of technology transfer be limited insofar as possible. Kristin Shrader-Frechette, redirecting her study of risk assessment to transnational transfers of technology, argues that each person in the developed world has an obligation to help prevent abuses of ill-monitored harmful technology transfers to Third World countries. Both Joseph Agassi and Mario Bunge call for global structure to facilitate dealing with global problems; Agassi argues for an international income tax in lieu of "foreign aid" as a way to transfer wealth from developed to developing nations; Bunge, for an international agency to manage the endangered biosphere.

These issues, one and all, are surely worthy of consideration. But, one might wonder, are they appropriate subject matter for philosophical reflection? Merely to leave the answer to this question to our readers is too facile, since our readers are selectively favorable to an affirmative. Alternatively, then, we would suggest that each contributor in his or her own way answers the question, albeit at times only implicitly. Our own perspective? The world is too important to leave to philosophers. But philosophy is too important to allow it to be carried on indefinitely without reference to the world in all its complexity.

Our thanks, finally, to Paul T. Durbin, general editor of this series, to all who participated in the review of the papers, to the conference participants, to those whose papers are being published in this volume, to the administration of Virginia Polytechnic for its gracious hospitality, and to all who have helped in the preparation of the volume, including Terry Mills in Indianapolis and Vanessa Alexander and Christine Duncan in Blacksburg.

Edmund F. Byrne
Indiana University
Indianapolis, Indiana

Joseph C. Pitt
Virginia Polytechnic Institute & State University
Blacksburg, Virginia

JOSEPH MARGOLIS

THE TECHNOLOGICAL SELF

I

There is a double puzzle that Thomas Kuhn collects in certain well-known remarks in his *The Structure of Scientific Revolutions* that compellingly links the theory of science and the theory of human inquiry – in effect, the theory of cognizing agents, of selves, of persons. One may doubt that Kuhn has formed an entirely coherent picture of the sciences, but there can be no question that he has completely neglected the analysis of what a human being must be like in order to live and work in the world he posits. Kuhn's linking these two issues remains instructive, nevertheless. For he grasps its paradoxical features in a way that does not really depend on the validity of his own account of the historicized sciences; and what he does say about the sciences is quite compatible with (indeed, it memorably instantiates) a number of very large doctrines that the entire sweep of Western philosophy may fairly now be said to be converging upon. These include at least: (a) the rejection of all forms of cognitive transparency and privilege; (b) the indissoluble unity of realist and idealist elements in any plausible theory of the sciences; (c) the conceptual symbiosis of cognizing self and cognized world; and (d) the matched historicity of self, science, and world.[1] Doctrines (a)–(d) dissolve any hierarchical advantage that might otherwise be assigned so-called naturalistic and phenomenological theories *vis-à-vis* one another and fix at the same time the sense in which theories of either sort could incorporate so-called deconstructive or post-structuralist exposés of their own pretensions regarding any form of cognitive transparency. By a term of art – a fair term – contemporary views incorporating (a)–(d) may be dubbed *pragmatist*.[2]

Kuhn's remarks are these: first of all, that "Lavoisier . . . saw oxygen where Priestley had seen dephlogisticated air and where others had seen nothing at all. . . . Lavoisier saw nature differently . . . Lavoisier worked in a different world";[3] secondly, speaking of that phase of post-fourteenth-century physics (affecting Galileo's work) in which Buridan and Oresme's impetus theory replaces Aristotle's, that "I"

1

Edmund F. Byrne and Joseph C. Pitt (eds), Technological Transformation: Contextual and Conceptual Implications, 1–15.
© *1989 Kluwer Academic Publishers.*

[that is, Kuhn] am . . . acutely aware of the difficulties created by saying that when Aristotle and Galileo looked at swinging stones, the first saw constrained fall, the second a pendulum."[4] Kuhn, of course, favors the thesis that these paired scientists "pursued their research in different worlds":

Until [for example] that scholastic paradigm was invented [Kuhn says], there were no pendulums, but only swinging stones, for the scientist to see. Pendulums were brought into existence by something very like a paradigm-induced gestalt switch.[5]

We are not interested here in the bafflements of Kuhn's own conception of the sciences except as they may help us to understand what is required of a theory of the cognitively apt selves that pursue particular inquiries under the conditions Kuhn advances or, more generally, under constraints (a)–(d) that Kuhn's own views instantiate. Kuhn gladly abandons all talk of "'the given' of experience," "immediate experience," "a pure observation-language," "mere neutral and objective reports on 'the given.'"[6] But he effectively reneges on this proviso – however unwittingly – in his explanation of the viability of the contingently different "worlds" of different societies: "An appropriately programmed perceptual mechanism," Kuhn explains, "has survival value. To say that the members of different groups may have different perceptions *when confronted with the same stimuli* is not to imply that they may have just any perceptions at all."[7] The remark is fair enough. But on what grounds (accessible to Kuhn) can we speak of the operations of "the same stimuli" across different paradigms, differently "programmed perceptual mechanisms"? "Two groups," Kuhn maintains,

the members of which have systematically different sensations on receipt of the same stimuli, do *in some sense* live in different worlds. We posit the existence of stimuli to explain our perceptions of the world, and we posit their immutability to avoid both individual and social solipsism. About neither posit have I the slightest reservation. But our world is populated in the first instance not by stimuli but by the objects of our sensations, and these need not be the same, individual to individual or group to group. To the extent, of course, that individuals belong to the same group and thus share education, language, experience, and culture, we have good reason to suppose that their sensations are the same. . . . They must see things, process stimuli, in much the same ways. But where the differentiation and specialization of groups begins, we have no similar evidence for the immutability of sensations.[8]

These are very curious remarks: first, because "invariance" or "immutability" of "stimuli" (neurophysiological connections, even physical laws) are merely *posited* to forestall solipsism (skepticism, radical incommensurability, intellectual nihilism, anarchy, relativism); second, because such invariances are themselves validly relativized to the shared "form of life" of a given society *and only there*; and third, because, apparently both intra- and intersocietally, the division of labor and historical variation *threaten our confirming any genuine, context-free invariances*.

Kuhn is not content with this kind of tenuousness. "We try," he says,

to interpret sensations already at hand, to analyze what is for us the given. However we do that, the processes involved must ultimately be neural, and they are therefore governed by the same *physico-chemical* laws that govern perception on the one hand and the beating of our hearts on the other. But the fact that the system obeys the same laws [in all perceptual cases, presumably in all societies] provides no reason to suppose that our neural apparatus is programmed to operate the same way in interpretation as in perception or in either as in the beating of our hearts.[9]

It is in this same context that Kuhn concludes that "An appropriately programmed perceptual mechanism has survival value."[9] This means that those who live in "different worlds" also live in "one world," that the provisional invariances internal to the different worlds of socially shared practices are also good guesses of some sort regarding the actual invariances that hold across such different worlds, that the "incommensurable viewpoints"[10] of these separate worlds are also collected within the range of commensurability (or, at least within the range of intelligibility) of the one overarching world. Incommensurability is not – or at least should not be – construed as equivalent to incommunicability or unintelligibility or untranslatability; on the contrary, moderate incommensurabilities, as much of conceptual categories as of metrical instruments, must, on pain of incoherence, be intelligible, even comparable, to the same inquirer or inquirers.[11] And yet, of course, *to be able to affirm invariances across moderate incommensurabilities signifies cognitive sources that cannot be confined within the bounds of such incommensurabilities.* Kuhn never explains that ability.

There is no question that Kuhn has put his finger on the essential puzzle of a historicized conception of science still bent on formulating the lawlike invariances of the entire order of physical nature. But it is equally clear that Kuhn's solution is threatened with an ineliminable measure of incoherence. For our present purpose, it is more important to emphasize what may be called the "constructive" or "constitutive" theme in Kuhn's theories, the notion that the world we live in – we ordinary percipients as well as Aristotle and Galileo as more disciplined scientists – is in some way *constituted* by the socially shared paradigms or practices that form or preform (tacitly rather than by explicit conjecture) the way we perceive and think. Kuhn sees the matter more in terms of the general nature and psychology of human investigators than in terms of the merely formal features of potential truth-claims advanced within the relevant space; and yet, he nowhere directly considers *what* a human person must be like – constituted and reconstituted *by* such cultural forces in the same instant in which the "world" is constituted and reconstituted *by* our changing inquiries and interventions. In this sense, Kuhn offers the barest glimpse of the interesting notion (which his own theory requires and which is required by any generic theory that subscribes to (a)–(d)): that *the human self is itself technologically and praxically constituted.* The potentially radical implications of this notion normally escape our notice, in spite of the fact that constraints (a)–(d) – perhaps, now, only marginally clarified by Kuhn's own favored theories – must surely be among the most salient conceded in our own age. The point may be taken as embedded at least in Kuhn's challenging distinction between a swinging stone and a pendulum.

We are marking off a strategy of argument, possibly a map of an argument, not an actual argument. The approach enjoys a considerable economy. For, there are a surprising number of quite powerful consequences that follow from admitting (a)–(d) together with the cognate finding that if "worlds" are constituted by the inquiries and practices of human selves, then selves are correspondingly constituted by processes internal to the formed worlds in which they contingently mature. The question of whether selves and persons may be eliminated by some ontological maneuver may be safely set aside: there is no known argument that actually effects that economy once we concede the reality of psychological experience (in however narrow or broad a sense we favor) or once we concede cognizing activities or actions informed by experience;[12] and, in any case, the eliminative maneuver can hardly pretend to have come to grips with the kind of puzzle Kuhn's examination of our historicized sciences imposes upon us.

Merely to concede the point of what may now be called (e), the thesis of the technological or technologized self, leads directly to a number of important findings – in a remarkably painless way. It affords a very simple conceptual lever by which to topple a large number of fashionable theories. For example, it follows instantly from the theory of the praxical or technic constitution of the self that *all* would-be findings of invariances, natural necessities, nomic universals, essences, closed systems, indubitability, self-evidence and the like *must be no more than idealized posits made within the indefinable limits of the competence and horizon of contingently formed and focused selves.* This is not to deny that it is entirely plausible to posit candidate regularities of such sorts. Indeed, we cannot avoid doing so if we are to accommodate (as we must) the achievement of science. First of all, mere empirical or Humean regularities – human mortality, for instance, or the fixity of the genetic code of the human species, or the ubiquity of certain anthropological constants[13] – hold only within the scope of those benign invariances Kuhn posits; and secondly, logical necessities, as Quine as shown,[14] can always be treated as open to being denied for reasons of systematic advantage (even if, characteristically, they do not actually require that we seize that advantage).

Alternatively put, it is not the indicative invariances of empirical science or of general human practice or of the (empirical) work of logic and mathematics that count as conceptually troublesome; it is only such invariances construed as entailing or presupposing some more profoundly fixed, universal, necessary, or apodictic invariances of *cogniscence.* The theory of the technologized self is, primarily, a theory of the contingently constituted, societally formed, historicized, diachronically alterable practices of actual human communities. Hence, the provisionality of any would-be universals – tendered in the spirit of Kuhn himself – may readily be allowed to stand. But there remain an endless number of imposing claims of considerable influence that either utterly fail to come to grips with the flat challenge of that thesis or else, more paradoxically, appear to embrace a version of it at the same time they insist on invariances that the thesis would clearly disallow. Thus, at a stroke, *if* the validity of induction and the analysis of the structure of a natural language in extensional terms were relativized to some "corpus of knowledge" (if they were treated as "epistemological" notions in that sense[15]), then inductivism and extensionalism would at best be no more than idealized presumptions that would never ensure their own preferability on logical or conceptual grounds – they could never be more than useful empirical projects.

In that sense, Peircean assurances about the long run, Popperian verisimilitude, Reichenbach's inductivism, Quine's repudiation of intentionality, Chomsky's nativism, Davidson's apriorism extending a Tarskian conception of truth to natural languages must all be abandoned at a stroke – which is hardly to deny that, *for limited runs of cases under the control of factors outside their own scope, such projects may indeed achieve a measure of success.*[16] To risk a term of opprobrium well beyond its usual application: all such views may be fairly tagged as *logocentric.*[17] Some, for instance Quine's, are actually (or ambivalently) sympathetic to minimizing innate invariances among human "subjects" (or selves): where Quine posits such fixities, he does so always *en bloc*, never distributively; and in doing that, he deliberately draws attention to no more than a principled continuity between the features of diverse cultures emerging from and embedded in a relatively uniform biology and the presumed fixities of that biology itself.[18] Quine's own attack on the analytic/synthetic distinction forbids specifying (as he himself explicitly observes) particular such fixities (in what we are calling the logocentric sense). Where such fixities are deliberately posited for the cognizing subject (the self or ego or whatnot), as in the transcendental gymnastics of Edmund Husserl's Ego or in the foundationalist presumptions of Roderick Chisholm's "first person", or where deliberate provision is made for such detailed disclosures, the thesis of the technologized self is instantly devastating – not, of course, in denying the contingent certainty of particular self-referential discoveries, only in subverting the presumption of their timeless inviolability or apodictic standing.[19]

The argument, then, may be put in a word: the world is a flux, reconstituted again and again through powers internal to its own contingent order, centered in local interventions at particular cognizing nodes. Correspondingly, the generic flux itself infects those provisionally specific nodes.

This single theme, economically summarized in (a)–(d) and brought to bear on the puzzle of (e), the constituted or constructed self, threads a disarmingly transparent path through much of the conceptual country of the most vigorously debated philosophies. But it is as instantly subversive as it is transparent. For it indicates how characteristically standard pretensions of universal fixities or even traditionally reliable regularities have presupposed the fixity of the self. What would remain, for instance, of the assured, reflexive, cognitive "beginnings" in the self, whether psychologically managed in the manner of Brentano or Chisholm or (more desperately) of Moritz Schlick[20] or egologically in the manner of Husserl, once the self, the locus of cognition, were admitted to be subject to the contingent vagaries of cultural construction? It is the presumption of the fixed nature of the self – the transformation (we may say) of the self into the immortal soul – that ensures the steady achievement of all familiar forms of foundationalism and the quest for the apodictic. That is the point of the original triumph of the Cartesian, well before the schism between naturalistic and phenomeno-logical strategies was invented. Challenge the assured unity of the self over any temporal span of self-consciousness, challenge the fixed, self-referential disclosures of *what* the self-cognizing powers of the self are: the epistemic certainty of any particular "self-presenting states" would fall into fundamental question.[21] Without that assurance, analytic epistemologies become completely vacuous, entirely formal. They lack legitimating concerns addressed to the praxical sources from which they emerge.[22] The intended assurance is incompatible with every theory that views the self as constructed or societally constituted or subject to historicized variability. The logical need for a referentially individuated center of cognition is simply not equivalent to any substantive doctrine regarding the actual epistemic or ontological unity or cognizing power of any such center: the *unicity* of the self is hardly equivalent to its supposed *unity* and hardly entails an original apodictic power.[23]

Once this is clear, one sees remarkably easily that any constructive conception of the world – roughly, within the purview of doctrines (b) and (c) – similarly undermines all hope of achieving a uniquely correct, or essentialist, or universalistic, or even verisimilitudinous science. We may suggest another powerful benefit of this line of argument (to which we shall have to return by way of a fresh start). *If* there is anything to the constructive theory of the self, then to that extent all theories of the structure of human history – of the narrative form of actual human existence and of the texts of authentic histories and fictionalized surro gates – cannot presume any essentialist or traditionalist structure, can-not for example (as with Paul Ricoeur) presume that Aristotle's *Poetics* and St. Augustine's *Confessions*, Book XI *have* captured the abiding ontological structure of man's historical existence.[24] That, too, would transform the self into a simple and timelessly invariant soul. But these remarks are meant only as reminders of the strategic and ubiquitous effect of theorizing about the self.

The critical finding on which these remarks converge is simply that it remains entirely open to us to search for the universal structures of the cognized world or of the cognizing mind, without presuming thereby to confirm the validity of any universalism – or, by parity of reasoning, any foundationalism, cognitivism, essentialism, inductivism, extensionalism and the like. Indicative first-order universals are entirely compatible

with the rejection of self-confirming second-order universals that would
fix our ultimate cognitive powers. The very forms of human tradition
and human culture are too various (and have always been too various)
to ensure any single, species-wide, convergent picture of the cognized
world and the cognizing self; but that variety has not actually disabled
the achievement of the sciences. Similarly, the constitution of the
cognizing self through the variety of cultures in which it emerges cannot
but subvert any reflexive pretension at marking the essential, innate,
species-wide structure of the mind's way of functioning; but that concession cannot meaningfully preclude the role of cognition itself. The
symbiosis of self and world is as neutral a posit as man seems capable of
grasping, once we concede our salient sciences. Hence, even the pretense of *not* requiring an account of the methodological and epistemic
puzzles of man's cognitive aptitudes – famously, in Hans-Georg Gadamer's
retreat from the false models of "truth" and "method" he would dismiss
– fails to escape a version of those same false models, fails to dislodge
the need for a theory of world and self. Gadamer's retreat is prophetic
of much of our philosophical age: the practice of the classical tradition
of Greece, either by itself or by analogy, is somehow made (by Gadamer) to confirm the cognizing norms of the race and is made to do
that without the need for any theoretical model of a methodological or
epistemic sort. The truth is: it abandons any such inquiry because it
ensures by "tradition," preemptively, what such an inquiry would
otherwise have had to establish.[25]

II

What we have sketched thus far are the lines of an argument by which,
admitting the constructive nature of the world along the moderate (if
somewhat muddled) lines of Kuhn's historicizing, we find ourselves
obliged to admit the constructive nature of cognizing selves. Mark that
(the constructive thesis) as thesis (1) of what we have termed the
doctrine of the technological or technologized self. It exercises an
immense economy in disqualifying at a stroke all forms of logocentrism
– all essentialisms, all universalisms, all natural necessities of cognition,
all totalizing, all closed systems, all apodicticity. But it is itself fragile
and incomplete as an account of what the technologized self entails. It

does not sufficiently identify what, minimally, the achievement of human communication requires.

If we allow the proliferation of something like very thin Wittgensteinian forms of life to function unchecked, "conventionally," anarchically, abstractly, without detailed constraints of any sort except of whatever is said to be internal to the constituting "rules" of this or that life form, we should make an utter mystery of effective human traditions, of the viable plurality of cultures, of the relative smoothness of inter-societal (as well as intra-societal) communication. We have already taken notice of Kuhn's rather naive effort to save – within the discontinuities of paradigm shifts – a certain fallback to an invariant physical order somehow not affected by that shift itself. *There* lies the incoherence threatening every historicized picture of science. It has already been energetically promoted in an anarchistic, even a deconstructive, spirit by Paul Feyerabend.[26]

The better clue lies elsewhere – in the biologized philosophical anthropologies of the European tradition. Marjorie Grene, for instance, captures what we shall mark here as theme (2) of the technologized self:

> To be a person is to be a history. In what respects? In two respects, opposed but related. On the one hand, being a person is an achievement of a living individual belonging to a natural kind whose genetic endowment and possible behaviors provide the necessary conditions for that achievement. On the other hand, a human being becomes the person he is within, and as one expression of, a complex network of artifacts – language, ritual, social institutions, styles of art and architecture, cosmologies and myths – that constitute a culture. A culture, of course, is itself a sedimentation of the actions of past persons; but it is, nevertheless, preexistent with respect to the development of any particular person.[27]

Speaking with Helmuth Plessner, Grene stresses "the natural artificiality of man" – that "it is our nature to need the artificial"; and, with Adolf Portmann, she speaks of the first year of life as "the year of the social uterus."[28] The biology of man uniquely requires completion through cultural artifactuality, through historicity (*Geschichtlichkeit*), through the development of the technologically apt self. Technology, then, is the biological aptitude of the human species for constituting, by alternative forms of equilibration, a world suited to a society of emergent selves or a society of such surviving selves adjusted, diachronically, to such a world. We understand one another for the same reason we survive as a species. Technology is the flowering of our biological endowment and is *incarnate* in it.[29]

This means that all communication, all understanding, all forms of knowledge, science, interpretation, action, cooperation, art, institutionalized life are biologically grounded, emergent with respect to – incarnations of – biological predispositions. Selves are constituted through the enculturing forms that our genetic endowment can support; and the diachronic reconstitution of selves accommodating successive such incarnations, generation by generation, must be: (i) open-ended toward future cultural forms: (ii) minimally constrained by the empirically projectible limits of innate capacities (short of evolutionary change); and (iii) similarly constrained by the empirically projectible limits of what the real structures of the world may be taken to be. These are all *holist* constraints – *not* privileged in any cognitive way, projected from *within* the "internalist" resources of a reflective science and philosophy.[30] They are, also, always addressed to reconciling the strong, *distributed* claims of our science with the deeper, *holist* sources of its realism. This is the only formula consistent with (our) theme (1) of the technological self and (2) of the need to account biologically for communicative successes of every sort.

It favors, therefore, two characteristic themes: first, the very strong doubt that the causalities of cultural process can be reduced to the terms of a closed physicalism of any sort or of a merely biologized system of closed generative rules – hence, it opposes the views of Donald Davidson and Noam Chomsky and Claude Lévi-Strauss; second, the very strong conviction that the underlying regularities, uniformities, similarities that make communication effective – as in the use of general linguistic terms – are themselves incarnate, technologized or praxical, resting on a tacit consensus about the tolerance of consensus, uncommitted to explicit universals, viable through the interpretive accommodations of an actual society.

III

We come, then, to our final theme. On the argument sketched, even the perception of a pendulum is a technological feat. There is no disjunction, at the cultural level, between thinking, perceiving, desiring, on the one hand, and acting, intervening, producing, inventing, on the other. The human self is a technologized organism, groomed in a natural way – that is, merely by living among the apt adults of a viable society – *to see* and think and act in accord with the Intentionally significant modes the various historical cultures have formed over time.[31] Hence, (3) theorizing is inherently praxical – in a double sense: first, because the very formation of the new selves of each new cohort of infants depends on processes of internalizing the culturally emergent forms of viable life a given society generates and because the diachronic, gradual reordering of the constitutive forms of each such society must be similarly congruent with the innate aptitudes that underlie every cultural evolution; and second, because the praxical achievement of man must be continuous with the sub-cultural aptitudes of other species, both instinctive and learned, that ensure their own viability under changing environmental conditions (notably, under conditions that man more and more extensively imposes). These themes lead to the *contextedness* of human life and to the *realism* due the cultural world. This is in part to say that, although the world is in a measure "constructed," it cannot be merely constructed (on pain of incoherence) and that its constructed phases must form a congenial and conservative set of biologically incarnate variations (on pain of excessive incommensurability).

The fact is that something analogous to the human situation must be admitted even among thoroughly instinctual animals. Something like the *ur*-form of the intentional unity of theory and *praxis* appears for instance in Niko Tinbergen's well-known studies of the stickleback. For, on the basis of Tinbergen's ingenious tests of the responsiveness of these aggressive fish to color spots and body shape and movement, it is quite clear that (in an instinctual way) they discriminate *abstract*, biologically pertinent universals; that is, they exhibit, *in their niches*, a behavioral responsiveness of an appropriate kind to runs of noticeably new physical stimuli *falling within approximate limits of some sort*.[32] Their behavior certainly invites intentional characterization (though Tinbergen would oppose the practice); and what they do within their niches would of course, in the human case (or among the higher animals), invite a fuller sense of intentionalized context.[33] The linguistic contexts of human intelligence preclude at one stroke both an extreme behaviorism and an extreme nominalism – since neither of those doctrines promises to analyze language itself in a completely satisfactory way.

The simple point remains that, among instinctual animals, there is no way to avoid postulating an internal information-processing mechanism of some sort that coordinates behavior and discrimination *within a range of invariance that clearly accommodates perceptual variances*. That is all

one means in speaking of "universals" among such creatures. By en-
larging the range of application from the animal to the human, by
rejecting any principled demarcation between the two, by raising universals to the status of a cultural achievement, by embedding learning in
a natural form of life, by linking the shifting limits of conceptual
tolerance and extension to intra-societal consensus and interpretation,
by binding the entire process to the embedding conditions of species
survival, by refusing to disjoin the conditions of effective thought and
perception and desire from those of effective action, by acknowledging
the informality of consensual implicatures of relevance and context, we
raise the primitive invariances of instinctual behavior to the conceptual
invariances of the technologized life of humans. There, they are simply
the benignly workable indicative invariances of any first-order empirical
inquiry, not the cognitively privileged invariances of any second-order
universalism or essentialism or the like. *The technological self is the
agent of that sort of aptitude.*

Furthermore, if we remind ourselves of Kuhn's constructive view of
reality (and concede thereupon the constructive nature of the self), we
see at once that we cannot, in a consistent way, construe selves realistically without so construing the "made" world as well. In particular,
since the emergence of the apt cognizing creatures we treat as persons
depends on the formative and sustaining powers of their particular
cultures, there is no plausible way in which to discredit the reality of
cultural phenomena. "To be a person is to be a history," Marjorie
Grene correctly affirms. To be a person is to be culturally constituted as
such. To be real as a person or as a self entails the reality of those
powers and attributes by which selves or persons are rendered real.
Nevertheless, in our own time as in others, there remains a very strong
tendency to confine the real to the merely physical – to movements
rather than actions, to living bodies rather than persons, to pigments
and blocks of stone rather than artworks, to strings of sound and strings
of marks rather than speech or language. We have already taken notice
of a certain similar tendency in Kuhn's worry about the relation between
swinging stones and pendulums. It reappears in Davidson's and in
Arthur Danto's much less troubled conceptions of action;[34] in the
elimination of persons or selves altogether in cognitive psychology;[35]
and in the tendency to treat the history, interpretation, and theory of art
(a metonym for the whole of human culture) as no more than heuristic
forms of rhetoric, favored *façons de parler* for viewing what, ultimately,

are no more than "mere real things" – reductively, physical objects.[36]

One cannot refuse the bare option of the reduction or elimination of the cultural dimension of the real. But its intended prize has yet to be earned. The doctrine of the technological self is incompatible with the victory of that project; and, in fact, the separate vindication of its own characteristic claims – the constructed nature of reality and self, the incarnation of cognition, the praxical nature of theory – counts against a bifurcation of the real and the rhetorical, in virtue of which one might be otherwise tempted to endorse their ultimate rejection. Failing that, we are invited to make a fresh analysis of what is clearly salient in human history – of what, in the opposing view, tends to be neglected anyway.

Nevertheless, in achieving just this small advantage, we have not yet explained what the sense is in which the technologized self or its world *are* constructed and yet are not *merely* constructed.

NOTES

[1] These issues are taken up in Joseph Margolis, *Science without Unity: Reconciling the Natural and Human Sciences* (Oxford: Basil Blackwell, 1987), Chs. 3–4.

[2] See Joseph Margolis, *Pragmatism without Foundations: Reconciling Realism and Relativism* (Oxford: Blackwell, 1986).

[3] Thomas S. Kuhn, *The Structure of Scientific Revolutions.*, 2nd ed. enl. (Chicago: University of Chicago Press, 1970), p. 118.

[4] *Ibid.*, p. 121.

[5] *Ibid.*, p. 120.

[6] *Ibid.*, pp. 126–127.

[7] *Ibid.*, p. 195; italics added.

[8] *Ibid.*, p. 193.

[9] *Ibid.*, p. 195.

[10] *Ibid.*, p. 200.

[11] The point is essentially missed by Donald Davidson. "The Very Idea of a Conceptual Scheme," *Inquiries into Truth and Interpretation* (Oxford: Clarendon, 1984). The mistake is of the greatest importance.

[12] The detailed argument is given in *Science without Unity*.

[13] See Clyde Kluckhohn *et al.*, "Values and Value-Orientation in the Theory of Action," Talcott Parsons and Edward A. Shils (eds.), *Toward a General Theory of Action* (Cambridge: Harvard University Press. 1951); and Florence Kluckhohn and Fred L. Strodbeck. *Variations in Value Orientations* (Evanston: Row, Peterson, 1981).

[14] This may be taken to be part of the force of W.V. Quine's "Two Dogmas of Empiricism," *From a Logical Point of View* (Cambridge: Harvard University Press, 1953). Quine's explicit remarks on this issue may be found in his *Word and Object* (Cambridge: MIT Press, 1960), pp. 11–12, 59–61.

14 JOSEPH MARGOLIS

See Isaac Levi, *The Enterprise of Knowledge* (Cambridge: MIT Press, 1980), p. 375. Levi's own theory of knowledge and inquiry attempts to preserve a measure of methodological invariance while admitting our dependence, contextually, on a changing corpus of knowledge. It also entails a very curious view of "epistemological infallibility." Chs. 1–3, for instance at pp. 67–68: also p. 13. See, also, the exchange on Isaac Levi's "Truth, Fallibility and the Growth of Knowledge," involving Israel Scheider and Avishai Margalit, in Robert S. Cohen and Marx Wartofsky (eds.), *Language, Logic and Method* (Dordrecht: D. Reidel, 1983).

For a specimen version of inductivism, see Hans Reichenbach, *Experience and Reduction* (Berkeley: University of California Press, 1938); see, also, Karl R. Popper, *Realism and the Aim of Science*, ed. W.W. Bartley, III (Totowa, N.J.: Rowman and Littlefield, 1983). Part I, Ch. 1. For a specimen of extensionalism, see Donald Davidson, *Inquiries into Truth and Interpretation* (Oxford: Clarendon, 1984); see, also, Ian Hacking, *Why Does Language Matter to Philosophy?* (Cambridge University Press, 1975), Ch. 12. Some specimens of the programs in question are examined in some depth in *Pragmatism without Foundations* and *Science without Unity*.

See Jacques Derrida, *Of Grammatology*, trans. Gayatri Spivak Chakravorty (Baltimore: Johns Hopkins University Press, 1976).

See W.V. Quine, *The Roots of Reference* (LaSalle, Ill.: Open Court, 1974), pp. 20–24.

See, for example, Edmund Husserl, *The Crisis of European Sciences and Transcendental Phenomenology*, trans. David Carr (Evanston: Northwestern University Press, 1970), § 54; and Roderick M. Chisholm, *The First Person* (Minneapolis: University of Minnesota Press, 1981), Ch. 7.

See Moritz Schlick, "The Foundation of Knowledge," trans. David Rynin, in A.J. Ayer (ed.), *Logical Positivism*, (Glencoe, Ill.: Free Press. 1959). Schlick excludes the possibility of error associated with protocol sentences (under Otto Neurath's barrage) by insisting that we are capable of cognizing "the absolute fixed points" of empirical science – given by what Schlick came to call an "observation statement" (as opposed to a "genuine protocol statement"), "always of the form 'Here now so and so' . . . because in a certain sense they cannot be written down at all." pp. 221, 223, 225.

The most subtle contemporary exploration of this question designed to salvage what can be salvaged from the Cartesian *cogito* may well be that of Roderick M. Chisholm. *First Person* (Minneapolis: University of Minnesota, 1981), p. 90. See, also, Hector-Neri Castañeda, "He: A Study in the Logic of Self-Consciousness," *Ratio*, VIII (1966); Elizabeth Anscombe, "The First Person," in Samuel Guttenplan (ed.). *Mind and Language* (Oxford: Clarendon, 1975).

The argument is given in *Pragmatism without Foundations*, Ch. 10. Cf. also, Alvin I. Goldman, *Epistemology and Cognition* (Cambridge: Harvard University Press, 1986).

See *Science without Unity*, Ch. 3.

See Paul Ricoeur, *Time and Narrative*, Vols. 1 and 2, trans. Kathleen McLaughlin and David Pellauer (Chicago: University of Chicago Press, 1984, 1985).

Apart from Gadamer, one may mention Taylor's and MacIntyre's views as specimens of traditionalism. See, for instance, Charles Taylor, "Philosophy and Its History," in Richard Rorty *et al.* (eds.), *Philosophy in History* (Cambridge: Cambridge University Press, 1984); and Alasdair MacIntyre, *After Virtue*, 2nd ed. (Notre Dame: Notre Dame University Press, 1984).

See Paul K. Feyerabend, *Against Method* (London: New Left Books, 1975).

[27] Marjorie Grene, "The Paradoxes of Historicity," in Brice R. Wachterhauser, *Hermeneutics and Modern Philosophy*, (Albany: SUNY Press, 1986) pp. 168–169.

[28] *Ibid*, p. 169. Cf. Marjorie Grene, *The Understanding of Nature* (Dordrecht: D. Reidel, 1974).

[29] For an account of the "incarnate," see *Science without Unity*, Ch. 9; Joseph Margolis, *Culture and Cultural Entities* (Dordrecht: D. Reidel, 1984), Ch. 1.

[30] See *Pragmatism without Foundations*, Ch. 11.

[31] For a fuller sense of the use of the term "Intentional" $=_{df}$ "cultural", see *Science without Unity*, Chs. 7, 9.

[32] See N. Tinbergen, *A Study of Instinct*, with a new introduction (New York: Oxford University Press, 1969); and *Culture and Cultural Entities*, Ch. 3.

[33] See David Premack, *Gavagail* (Cambridge: MIT Press 1986).

[34] Donald Davidson, "Agency," *Essays on Actions and Events* (Oxford: Clarendon, 1980); Arthur C. Danto, *Analytical Philosophy of Action* (Cambridge: Cambridge University Press, 1973.

[35] See for instance Stephen P. Stich, *From Folk Psychology to Cognitive Science* (Cambridge: MIT Press, 1983); also, *Science without Unity*, Ch. 5.

[36] The latter theme is pursued most systematically by Danto. See his *Transfiguration of the Commonplace* (Cambridge: Harvard University Press, 1981); and *The Philosophical Disenfranchisement of Art* (New York: Columbia University Press, 1986). I have examined Danto's account in "Ontology Down and Out in Art and Science," forthcoming in *Journal of Aesthetics and Art Criticism*. Cf., also; Joseph Margolis, *Art and Philosophy* (Atlantic Highlands, N.J.: Humanities Press, 1980).

TYRONE LAI

CRYPTANALYSIS: UNCOVERING OBJECTIVE KNOWLEDGE OF HIDDEN REALITIES

THE UNCOVERING OF OBJECTIVE KNOWLEDGE OF HIDDEN REALITIES IS MADE POSSIBLE BY AN OPERATING PRINCIPLE

In this paper, using cryptanalysis as an example, I explain a general method, or strategy, already in use, for uncovering objective knowledge of hidden realities. I call this strategy IVS (the investigative strategy). To help in the understanding of the IVS process, I shall compare IVS, itself an operating principle, with other operating principles (I shall explain what an operating principle is in the body of the paper). Two points, among others, I shall be addressing myself to are the following.

1. Why the search for objective knowledge is possible even when evidence is always interpreted in the light of the theory it 'supports.'
2. How the misunderstanding, or paradox – the Meno paradox, arises that it is in principle impossible, and therefore irrational to try, to uncover objective knowledge of hidden realities.

CRYPTANALYSIS AND ITS TWO CENTRAL PROBLEMS

In cryptology (for an introduction to the subject, see Kahn (1967), we distinguish between a cryptogram, its cleartext, and the cipher. From the cryptogram, with the cipher, we can deduce the cleartext (decipherment or decoding). From the cleartext, with the cipher, we can deduce (generate) the cryptogram (encipherment or encoding). If we are only given the cryptogram and are asked to find the cipher and the cleartext, we have to resort to cryptanalysis.

Cryptanalysis is a form of investigation directed at the uncovering of objective knowledge. Clearly, it is the actual cipher and the actual cleartext we want, not a cipher and a cleartext we ourselves conjure up.

Cryptanalysis is not a deductive process. The knowledge we uncover through cryptanalysis is new knowledge, knowledge which we originally did not have. The reason that cryptanalysis is not a deductive process is

17

Edmund F. Byrne and Joseph C. Pitt (eds), Technological Transformation: Contextual and Conceptual Implications, 17–32.
© *1989 Kluwer Academic Publishers.*

very simply that from the cryptogram we cannot deduce both the cipher and the cleartext; there are not enough premises. Encipherment and decipherment are deductive processes, as we have pointed out, but in both encipherment and decipherment, the cipher is given and can thus be included in the premises.

Some may say, from the cryptogram together with our background knowledge, we can deduce both the cipher and the cleartext. To this the answer is, our background knowledge is vast. We cannot try out every single item to see which, together with the cryptogram, will lead deductively to the cipher and the cleartext. Thus, *even if* the cipher and the cleartext can be deduced from the cryptogram and our background knowledge, we have to *discover* which items of our background knowledge will enable us to carry out this deduction. This attempted discovery, when successful, is discovery of knowledge which we originally did not have.

Now, it is true that in trying to discover which items of our background knowledge is useful, clues will help. But clues too have to be discovered! They are not given. True too, some clues are easy to find, but others are not. Naturally, cryptographers (those who make up ciphers) will try not to leave obvious clues. If they can, they will try not to leave any clues at all.

Some clues, when they are discovered, alert us to items in our background knowledge, but this is not the only thing clues do. *Some clues lead to new knowledge.* This is a common phenomenon. We learn, for example, new words from context. Context, in this case, acts as the clue. In view of this possibility, cryptanalysts are able to decipher (cryptanalyse) dead languages. If most of a dead language has rough parallels in related languages, the residue can be deciphered from context. *Cryptanalysis is not limited to languages we already know*; this is important to keep in mind.

In cryptanalysis, we do not have direct access to the realities of which we want knowledge: the cipher and the cleartext are locked up somewhere, in a desk, in a vault, in some one person's head. . . . To uncover knowledge of these realities, we face two main problems: how to proceed and how to evaluate results. The second of these two problems is the more intriguing. Since the realities we want to know are hidden, we cannot evaluate results by comparing them directly with these hidden realities. But if we cannot evaluate results through direct comparison, how can we ensure that our results are objectively right!

This second problem appears formidable. But, as we all know, it is soluble. As is the problem about procedure. Since cryptanalysis is in fact possible, at least sometimes.

That cryptanalysis is possible shows us that *it is possible to uncover objective knowledge of hidden realities*. The fact that the realities we are interested in are hidden; the fact that we do not have direct access to them; the fact that we cannot evaluate results through direct comparison; does not mean that we may not have some degree of success when trying to uncover this kind of knowledge.

That cryptanalysis is possible – that is, humanly possible – is nowadays well recognized. At one time, it was not the case. In the Middle Ages, cryptanalysts were thought to be in league with the devil (Kahn (1967)). And not just by the uneducated! How else, it was asked, could cryptanalysts tell they had gotten the real message – when the latter was purposely hidden from them?

Now, while we all know that cryptanalysis is possible, we should also be aware that success in cryptanalysis is not guaranteed. There is a general strategy we can follow, it is true, but other things are also needed; for example, ingenuity. When some of these other things are missing, we may fail, or we succeed partially (for, an interesting, real-life, partial cryptanalysis, see Brumbaugh (1978)). And, of course, we are never absolutely certain of our results.

A TRIADIC RELATION

In cryptanalysis, we are concerned with a triadic relation. There are three structures, or patterns, of interest to us: the cryptogram, the cipher, and the cleartext. These three patterns are related: from the cryptogram and the cipher, we should be able to deduce the cleartext; from the cleartext and the cipher, we should be able to deduce the cryptogram. In cryptanalysis, two of these three patterns are missing: the cipher and the cleartext. Sometimes even parts, or the whole, of the third, the cryptogram (encoded messages are sometimes buried inside continuous random radio transmission). In cryptanalysis, therefore, we look for at least two things. Sometimes we look for three. And we have to do so without being able to verify our results directly.

Now, if we call the cryptogram the evidence, the cleartext, the interpretation of evidence, and the cipher recovered, the theory, we can then notice that cryptanalysis is not just a matter of finding an interpretation of the evidence, but a matter of finding an interpretation of the evidence in accordance with the theory advanced.

We should also notice that in this triadic relation, no assumption is made that the evidence, the cryptogram, is theory-free; it could have been transcribed from Morse, for example; or a telegraphic code different from Morse, which has to be cryptanalysed first.

Also, in those cases where parts of the cryptogram are missing, the location of the missing parts may have to depend on the cipher in the process of being recovered. For example, suppose I have recovered most of a cipher, but the cryptogram I have is not sufficient to enable me to recover the rest. Now I have come into possession of a new cryptogram. If it is from the same cipher, it may enable me to recover the rest of the cipher. But how do I know it is from the same cipher? Clearly, the way to find out is to see how much of the new cryptogram can be deciphered by the part of the cipher I already know.

The search for evidence, even in cryptanalysis, is guided by theories.

OTHER KINDS OF INVESTIGATIONS SIMILAR TO CRYPTANALYSIS

Cryptanalysis is not the only kind of investigation in which results cannot be directly verified. In other common kinds of investigations – common nowadays but not so common once upon a time – we have the same situation. Criminals do not always confess to their crimes. (In some countries, a person has a right to avoid self-incrimination.) There may be no eye-witnesses at the scene of the crime. Yet, despite all this, detectives are sometimes able to 'piece the story together'; that is, offer a credible theory as to how, when, where, why, and by whom, the crime was committed; and, based on this theory, an interpretation of the evidence gathered (for example, sharp object, blood-stained clothing, etc.); which evidence, interpreted according to the theory that has been advanced, they, the detectives, or the prosecuting attorneys, may have to present in court. In science, too, there is no direct verification of discoveries. When we discover a new theory in science, we interpret the evidence which 'supports' the theory *according to the theory*; there is no direct verification either of the theory or of the interpretation based on it. When Newton discovers the theory of universal gravitation, he interprets the planets as material bodies which obey his theory. Traditionally, planets were believed to be made of the fifth essence.

That evidence is interpreted in accordance with the theory it 'supports' often leads people to conclude that it is impossible to uncover objective knowledge. We shall see later why this conclusion does not follow.

OPERATING PRINCIPLES

How can we, in cryptanalysis, uncover both the cipher and the cleartext? *And* be able to tell whether we have done so? How do we carry out an investigation directed at uncovering objective, hidden, knowledge? How do we evaluate results in such an investigation?

We follow a strategy, or, we may say, an operating principle. We shall call this operating principle, this strategy, IVS.

What are operating principles?
They are principles which tell us in a general way how to arrange things and processes so that we are able to do things which, without them, we either cannot easily do, or cannot do at all. For example, there is the principle of the wheel: it enables us to transport heavy loads over rough surfaces with relatively little effort. There is also the principle of the airfoil: it allows us to fly without growing wings. IVS allows us to uncover objective knowledge without direct access to the realities of which we want knowledge.

Now if IVS does enable us to uncover knowledge of hidden realities, there must be, within IVS, a method of evaluating results *which does not depend on direct access* (to the realities in which we are interested). We shall explain what this method is in the next section.

We have accumulated a large number of operating principles by this time; civilization depends on them. IVS is one of these operating principles. IVS allows us to take on what appears to be a formidable, and to many, even an impossible, task.

IVS AND MPE

What exactly is IVS? What do we do when we follow IVS? How do we arrive at IVS? How do we evaluate results within IVS?

To answer these questions, let us turn to cryptanalysis again. In cryptanalysis, we have to arrive both at the cipher and the cleartext. Except for the simplest cases, the cipher and the cleartext are structures, patterns, of some complexity. Normally it is not likely that we are able to arrive at either one of them in one single step. We have to do things gradually, a little at a time. Our task has to be accomplished through a piecemeal process.

Now, even by a piecemeal process, we are not likely to be able to find the cipher first, and then the cleartext. Indeed, now that we have decided to carry out our task in a piecemeal fashion, there is no reason why we should look for the two – the cipher and the cleartext – separately. The cipher and the cleartext, we know, have to harmonize: from the cryptogram and the cipher, we should be able to deduce the cleartext. Because they have to harmonize, if we succeed in uncovering part of the cipher, we will have succeeded in uncovering part of the cleartext. This suggests that we should uncover the two of them in parallel.

Taking a parallel approach also has this advantage. To uncover the cipher, we need clues. Initially we are not likely to have many clues. However, if we succeed in uncovering part of the cipher, by applying this part of the cipher to the cryptogram, we may be able to develop new clues. New clues will enable us to uncover more of the cipher. Knowing more of the cipher will enable us to decipher more of the cryptogram, which may lead to more new clues. This means, if we are able to make a small successful beginning, by adopting a parallel approach, we have the chance of continuing on to the end, recovering eventually both the cipher and the cleartext. Now, on the way, we could make mistakes. But this is not serious. For, minor mistakes we can correct as we go along, as we uncover more. If we make serious mistakes, we still have a chance of finding out and correcting them. For, if we make serious mistakes in decipherment, we will not derive any new clues, or we derive false clues. With no new clues, it is impossible to uncover more of the cipher. If we have false clues, they will lead to more mistakes. Thus, if we make serious mistakes, either in uncovering the cipher, or in uncovering the cleartext, our investigation will, sooner or later, come to a stop, an impasse. At that point, we can retrace our steps and try to see if we can correct our mistakes. If we are able to correct them, we should be able to advance again. (Correcting mistakes this way is often a matter of trial-and-error. The indication that we have corrected the mistakes is that we are able to advance again.) Thus, by proceeding in a piecemeal fashion, and by engaging in two parallel, mutually fortifying, processes, one dedicated to the uncovering of the cipher, the other to the uncovering of the cleartext, we may be able eventually to uncover both. This then is IVS. In sum, IVS is the following. To uncover the hidden realities, we engage in two parallel processes and let them a) help each other forward and b) correct each other. If we are able to move forward, we are doing things right. If we reach an impasse, we are likely to have made mistakes. From this, we see that IVS contains within itself a method of evaluating results. We call this method MPE (the method of progressive evaluation).

Those who have had experience with investigations, I am sure, are familiar with both IVS and MPE. Those who have not and want to have some experience with IVS and MPE first hand, are asked to try the following simple exercise in cryptanalysis, in which a beginning has already been made (which is not to say, however, that the beginning is necessarily correct; for examples in cryptanalysis closer to real life, see Friedman (1976)).

Cleartext: SHE
Cryptogram: SBR SBCTU DBCKERVS FCGG WTTCXR SFH
                   FRRJD YTHE SHUWI

RELATIVISM?

In following IVS, we evaluate results by MPE; that is, by seeing whether we are moving forward with the investigation. If we are able to move forward, our results are likely to be right. Now, some may ask, but how do we know whether we are moving forward? Can people not have different ideas as to what moving forward, or progress, is? Will IVS not lead to relativism?

Now, in talking about IVS in the abstract, as we are doing here, it can easily appear that relativism is a real possibility. For, how we gauge whether we are making progress when following IVS depends on what assumptions we have made. Suppose we have made the assumption that the cleartext is in English. If after making this assumption, we are able to decipher more and more English words, we are then making progress. But suppose we have made the assumption that the cleartext is in

German, finding English words in the cleartext then will not lead us to conclude that we are making progress – so long as we keep to our assumption. So it may appear that progress is relative; relative to the assumptions we have made.

But this conclusion – that progress is relative – cannot be right. For, even though we have no direct access to the cleartext, we obviously cannot decide by fiat in what language the cleartext should be. If, somehow, we keep uncovering only English words in the cleartext, the cleartext then cannot be in German. The assumption that it is is simply false. The fact of the matter is, *in IVS we evaluate assumptions also by MPE. If we make wrong assumptions, we will not be able to make any progress.* If the cleartext is in English, it is impossible to impose a German message on the cryptogram. Any attempt on our part will lead to impasse *long before we finish deciphering the whole message.* (Our assumptions do not have to be exactly right; we can do with approximations. If we make the assumption that the cleartext is in English, meaning ordinary English, when in fact it is in Newfoundland English, it will not hurt; we can proceed and discover the mistake ourselves later on. We have already pointed out that through the use of rough parallels, we can decipher dead languages. See also the section on sleepwalkers later on in this paper.) The following is important to bring to mind again. In cryptanalysis, we are dealing with a triadic relation. There are three patterns we are interested in, not two: the cryptogram (or evidence), the cipher (or theory), and the cleartext (the interpretation of evidence according to theory). These three patterns have to harmonize. None of them is formless matter. As a result, in IVS progress is not unilinear – it is not a matter of giving form to more and more of matter which previously is without form; but zig-zagged – it is progress in two, and sometimes even three, mutually fortifying, processes. (There are three mutually fortifying processes when the cryptogram (the evidence) is not given, but has to be searched out.) We cannot counterfeit this kind of zig-zag progress.

Notice that we do not recognize this kind of zig-zag progress by some universal criterion which we establish separately; we recognize it, as we have explained, based on assumptions we make in each particular case, which assumptions however would not lead to progress if they were not right.

WHY IVS WORKS

Operating principles, when they work, do so for objective reasons. IVS works. Why? What are the objective reasons? We can now answer this question very simply. When IVS works, it works because there are objective, hidden, harmonizing patterns (plural) behind the evidence; the evidence, together with its interpretation and the theory, form a triadic relation. If there were no such hidden patterns; if there were no such triadic relation; IVS would not work. We cannot cryptanalyse a random string of symbols. (Cryptographers, to increase the security of messages transmitted by radio, sometimes sandwich encrypted messages between random transmissions.)

INVESTIGATIONS ON TOP OF INVESTIGATIONS

The effectiveness of IVS, and thus also of MPE, is increased when we stack investigations on top of other investigations. For example, a message retrieved through cryptanalysis could be (and often is) made into an ingredient of a larger investigation. The results of the larger investigation could be made into ingredients of an investigation larger still; and so on. In this way, when we make progress in all these investigations, we will be uncovering an expanding complex of harmonizing patterns, a complex which, difficult to counterfeit to start with, is progressively even more difficult to counterfeit.

NO CRITERION OF TRUTH

Now that we know why IVS works, we can clear up a common misunderstanding.

It is sometimes thought that in investigations, and especially in cryptanalysis, we evaluate results by a criterion of truth: coherence. Now, this can only be a misunderstanding. In investigations, we follow IVS. And we have seen already we do not evaluate results by a criterion of truth in IVS, but by MPE. How then does this misunderstanding arise? Why would people think there is a criterion of truth when there isn't? When the concept itself (of a criterion of truth) is known to be problematic?

The misunderstanding arises this way. IVS would not work unless there are hidden patterns (plural). In following IVS, we definitely look for patterns: if we did not suspect there were hidden patterns, we would not be engaging in the investigation. However, *in an investigation, we do not evaluate results at the end*, when everything is done, by seeing at that point whether the results are coherent. We evaluate results as we go along, by MPE. We have already explained why we have to proceed this way, in a piecemeal fashion. The reason is, the hidden patterns not only have to be coherent each within itself, they have also to harmonize! It is only by proceeding in a piecemeal fashion that we have a chance of finding these hidden harmonizing patterns. But this means we have to evaluate results as we go along. And this we do by MPE. Thus, even though at the end the results form themselves into coherent patterns, we do not use coherence as a criterion in evaluating results.

There is also this important point. Results in investigations are only interim; any investigation can in theory be merged with some wider investigation. This has the consequence that results of the wider investigation may require us to correct the results of earlier investigations. For example, in cryptanalysis, we might have arrived at the message, 'Mr. Green will be in town on the 8th.' After a more extensive investigation, we may conclude that this message, though certainly coherent within itself, is not the true message. There has been a transcription mistake: the '8th' should have been the '9th'!

DISADVANTAGE INTO ADVANTAGE

We pointed out earlier that IVS is an operating principle, and that as such, it enables us to do things which at first sight might seem impossible. That they enable us to do things which might appear impossible is one of the common characteristics of operating principles. There is another. Operating principles are often able to turn what appears to be disadvantages into advantages. Pulling a load over a rough surface wastes energy – due to the friction between the load and the rough surface. Friction appears to be a disadvantage. However, in using a wheel, we make use of friction. Friction allows the wheel to 'grip.' The wheel turns what appears to be disadvantage into advantage. Without friction, the wheel will slip. A similar situation occurs with the airfoil. Air seems to be a disadvantage when it comes to holding up weight.

However, by letting air flow over the airfoil, we generate lift.

In an investigation, we have to uncover up to three things at the same time: the evidence, its interpretation, and the theory. This at first sight appears as a disadvantage. It seems what we are required to do is to interpret the evidence in our own favour; that is, in favour of the theory we are advancing. But IVS turns this apparent disadvantage into an advantage. For, it is only by exploiting the relation between the three patterns – the evidence, its interpretation, and the theory – that IVS is able to work. If the three patterns were unconnected, we would not have been able to devise IVS.

Now, the exploitation of this triadic relation will not lead to stacking the cards in our own favour. For, in IVS, we do not evaluate theory by pointing to the fact that theory can explain (provide an interpretation for) evidence; we evaluate *both* theory *and* interpretation of evidence *as they emerge* by MPE; that is, by seeing whether they lead to progress in the investigation (see also Lai (1988)). This we have to remember, with MPE, we do not evaluate theory (for example, the cipher recovered), nor interpretation (the cleartext recovered), as a whole, but in a piecemeal fashion. Some parts of the theory; the 'earlier' parts; are more certain than the later ones. So are some parts of the interpretation. The uncovering of theory, and the uncovering of interpretation, are piecemeal. So is their evaluation.

There is also this point. Evaluation of theory by MPE is not a logical exercise. From a logical standpoint, it appears odd that we should interpret the evidence in accordance with the theory we are advancing. Odd in that we ordinarily think that we use this interpretation to logically 'support' our theory. But this common belief is false; we do not evaluate our theory this way. We evaluate *both* theory *and* interpretation of evidence by MPE. Evaluation by MPE is methodological, not logical. Theory and interpretation have to harmonize. *If they do not the investigation cannot continue*!

DIFFERENT OPERATING PRINCIPLES HAVE DIFFERENT PROPERTIES

This should be obvious. A wheel needs a surface for support if it is to function; it is quite useless when suspended in mid-air. An airfoil, on the other hand, needs airflow for it to work; it is useless in empty space. Wheels and wings are useful things but we would not for this reason attach them to everything; it depends on what we want done. We choose our operating principle according to our aim. Different operating principles have different properties. I draw attention to this point in order to remove certain confusions relating to the notion of rationality as it relates to the search for knowledge.

There is an operating principle, extremely simple, with which most of us are familiar. It does not, so far as I know, have a name. For the sake of convenience, I shall call it the building block principle, or BBP. Children, using building blocks, can build all kinds of structures following this principle. Housing contractors, following the same principle, can build all kinds of houses: it is a matter of putting the right pieces together in the right way. And this is what I mean by BBP. When we build a house, we follow a process in which steps are clearly laid down. When these steps are followed faithfully (assuming they have been properly designed), at the end of the process, we have the finished house.

In following BBP, there can be sub-processes within the main process. For example, doors and windows can be built separately. When they are ready, they can then be integrated into the main process.

BBP has its special properties. When we rely on BBP to reach an aim, we have to know in advance, and know clearly, *all the important details* of the aim. The reason for this is, there are infinitely many ways of putting together building blocks. If we are not clear what end-product we want, the chance is extremely high that we will have an end-product we don't want. In following BBP it is not sufficient merely to have a rough idea of the aim. For example, when we want to build a house, it is not sufficient merely to tell the contractor we want a house. It is not that he/she does not understand the word 'house,' but this information about our aim is not sufficiently detailed to tell him what he should do. Instead of merely telling the contractor we want a house, the usual practice is to provide him with a blueprint in which all the important details are specified. Such a blueprint performs two important functions. 1) It enables the contractor to know precisely what he should do. 2) It enables him to evaluate results; to determine whether he has succeeded in building the house agreed upon.

In following BBP, with simpler projects, we can sometimes avoid laying down specifications, or criteria (although we still have to be clear in advance what end-product we want). An experienced potter, after making one teacup, could easily make another copy of the same teacup.

In this case, we judge whether the two copies are sufficiently close simply by inspection.

This requirement, that we be clear as to the important details of our aims – either through direct acquaintance or in virtue of a set of criteria, which we find in BBP, let us call the clarity requirement.

We do not find the clarity requirement in IVS. In cryptanalysis, for example, we do not know in advance the important details of the end-product, or rather, the end-products: since there are at least two things we want: the cipher and the cleartext. If we knew, we would not have to resort to cryptanalysis. But, some may ask, if we do not know in advance the important details of the end-products, how can we rationally determine what we should do? And how are we, objectively, to evaluate results? Our answers to these questions are, in cryptanalysis, we follow IVS, and we evaluate results by MPE.

At this point, we can see the origin of the Meno paradox: the contention that it is irrational to search for objective knowledge of hidden realities. This is how the paradox appears in the *Meno*:

MENO: But how will you look for something when you don't in the least know what it is? How on earth are you going to set up something you don't know as the object of your search? To put it another way, even if you come right up against it, how will you know that what you have found is the thing you didn't know?

SOCRATES: I know what you mean. Do you realize that what you are bringing up is the trick argument that a man cannot try to discover either what he knows or what he does not know? He would not seek what he knows, for since he knows it there is no need for the inquiry, nor what he does not know, for in that case he does not even know what he is to look for.

We should now be able to see that this paradox is based on the mistaken assumption that the search for knowledge is a BBP. If the search for knowledge were a BBP, we would have to know in advance, and in detail, what it is we are searching for. But of course if we know already, there is no need for the search. And if we do not know, there will be no point for the search – since for one thing we will not be able to evaluate results. Hence if the search for knowledge is a BBP, it is irrational.

But the search for knowledge is not a BBP. To search for knowledge, we follow IVS. It is perfectly rational to search for knowledge without knowing in advance the important details of the knowledge we are searching for. The clarity requirement is not a part of IVS.

Now it is true that when we are not clear about the important details of our aims, we cannot be precise about the steps we take. *But this in practice we know.* In investigations, we know we have to rely on trial and error, and on approximations.

It should not be surprising that we do not find the clarity requirement in IVS. IVS is a different operating principle from BBP. Different operating principles have different properties.

SLEEPWALKERS

We do not follow BBP when we are searching for knowledge; we follow IVS. In doing so, we do not have to be clear about the details of our aims. Indeed, we are often hazy even as to a *rough* statement of our aims. For example, a person has disappeared. Should we start an investigation? What should we look for? The person? He might be dead. The body? He might have been incinerated. The murderers? It might have been suicide. Often it is only after we have started the investigation that we become clearer as to the *general* nature of our aims. For example, we might come to suspect, after the investigation has started, that the person who is said to have disappeared never existed. One of the aims of our investigation then becomes, why do so many people say he disappeared?

In investigations, sometimes people do not find what they are looking for, but discover instead what they are not looking for. Kepler was looking for a way to fit the planetary orbits into the five regular solids. He failed. Instead, he discovered his three laws of planetary motion. Koestler calls Kepler a sleepwalker.

Since in investigations we cannot be clear as to the details of our aims, we cannot be precise as to what we should do and accurate as to the hypotheses we put forward. Yet, this lack of precision and accuracy will not necessarily prevent the investigation from going forward. A lot of the work Kepler put in in his astronomical researches led nowhere. And he was only approximately right, for example, when he says planets move in ellipses. But approximate discoveries like this were sufficient for Newton to continue with the investigation which eventually led to his, Newton's, discoveries.

The presence of inaccuracies in investigations is no cause for alarm. For, if the investigation is able to advance; if, that is to say, we are able to discover more; we can turn back and correct mistakes. We can do this because when we know more, we have a fuller context. It is a common occurrence that we discover new things, *things that we did not know before*, from context. For example, the meanings of new words.

An investigation is a piecemeal process. It is a piecemeal process in which we are heavily dependent on taking small steps one or a few at a time. These steps do not have to be exactly right. If need be, we can start with approximations and then improve upon them in small steps.

That we are dependent on taking small steps in an investigation we may call the small step principle. It is useful to remind ourselves of this principle: it helps foster patience and persistence.

IVS IS LAB- AND FIELD-TESTED

Like other operating principles, IVS can be empirically tested, both in the 'laboratory' and in the field. To lab-test IVS, we cover up some knowledge and have others try to uncover it through IVS. This way, we can determine whether IVS is effective by checking the results derived through IVS against what we ourselves know. This we do, for example, in exercises in cryptanalysis. To field-test IVS, we apply it to knowledge to which we have no hope in having direct access, either in theory or in practice. The possibility of cryptanalysis in real life, the success of science, the occasional success in crime detection – instances such as these, seem to indicate that IVS has performed reasonably well in the field.

CONCLUDING REMARKS

The question how we uncover objective knowledge of hidden realities is a philosophical question. In this paper, in IVS, I have found an answer to this question. In that IVS is an operating principle, and in that operating principles usually fall within the purview of technology, what I have done may be described as providing a technological answer to a philosophical question. Some may find this abhorent: they may see it as another instance of technological imperialism, this time invading philosophy itself. I do not know what my chance is of being excused. My defence is that the question is important; those who love truth will not mind where the answer comes from.

REFERENCES

Brumbaugh, R.S. *The Most Mysterious Manuscript: The Voynich 'Roger Bacon' Cipher Manuscript* (Carbondale: Southern Illinois University Press, 1978).

Friedman, W.F. *Cryptography and Cryptanalysis Articles*, 2 vols. (Laguna Hills, California: Aegean Park Press, 1976).

Kahn, D. *The Codebreakers* (New York: Macmillan, 1967).

Lai, T. 'Empirical Tests Are Only Auxiliary Devices.' Forthcoming, *British Journal for the Philosophy of Science*, Vol. 39(1988), pp. 211–223.

Plato. *Meno*, trans. W.K.C. Guthrie, in *The Collected Dialogues of Plato* (Princeton: Princeton University Press, 1969).

PAUL T. DURBIN

RESEARCH AND DEVELOPMENT FROM THE VIEWPOINT OF SOCIAL PHILOSOPHY

From a logical point of view, the strength of the support that a hypothesis receives from a given body of data should depend only on what the hypothesis asserts and what the data are: . . . a purely historical matter . . . should not count as affecting the confirmation of the hypothesis.

Carl G. Hempel, *Philosophy of Natural Science*, p. 38

Why [should] the conduct of engineering/science in an R&D setting [differ] from the conduct of engineering/science in other settings [?] For that matter, why should we think that philosophy of engineering differs from philosophy of science in an R&D setting?

Anonymous Referee

In December, 1987, Charles J. Pedersen shared a Nobel prize in chemistry for the discovery of the crown ether molecule. Pedersen spent his entire career as a chemist in the industrial laboratories of the DuPont Company. His fellow Nobel laureates – Donald Cram of the University of California at Los Angeles and Jean-Marie Lehn of the Louis Pasteur University in Strasbourg, France – worked in much more traditional academic settings. They expanded on Pedersen's work, with Cram developing "host-guest" chemistry that may hold significant promise for understanding cell behavior in living organisms and Lehn developing "cryptand" chemistry that has led to dissolving some of the most insoluble substances known to chemistry.[1] What better example could there be that the social setting has no influence on scientific results? A pure science discovery can come out of an industrial laboratory, and practical results can be developed in an academic setting.

I believe that there are important differences that set off science and engineering in research and development settings from so-called "pure science." As a result, I further believe that we need a distinctive

33

Edmund F. Byrne and Joseph C. Pitt (eds), *Technological Transformation: Contextual and Conceptual Implications*, 33–45.

philosophy of R&D that is not simply an extrapolation from philosophy of science – even of the anti-positivist sort – or philosophy of engineering (as if there were anything of the latter sort).[2] This article represents only a fragment of a larger effort to defend this point of view.[3]

As I will show, recent developments in anti-positivist philosophy of science – the example notwithstanding – might suggest that, if philosophy of science were to be extended to cover R&D science, some major changes would be required. But I go further than this. I believe (a) that philosophers of technology have neglected engineering practice (so often exercised in and around R&D settings), and (b) that, if they turn to such practice, an adequate philosophy of engineering and science in R&D settings will inevitably be situation-specific. That is, contrary to Hempel, the historical setting does affect truth claims in R& D science and engineering.

1. ANTI-POSITIVIST PHILOSOPHY OF SCIENCE AND THE NO-DIFFERENCE THESIS

a. It has become a cliché to date the historicist/anti-positivist movement in philosophy of science from the publication of Thomas Kuhn's *The Structure of Scientific Revolutions* in 1962,[4] so I will bow to custom and also begin there. By now, twenty-five years later, the main thrust of Kuhn's attack on the positivism of both Carnapian inductivism and Popperian falsificationism is well known. Kuhn's conclusion – that he is describing a "process that should somehow, in a theory of scientific inquiry, replace the confirmation or falsification procedures made familiar by our usual image of science"[5] – is based on his reading of major revolutions in the history of the physical sciences from the Copernican revolution in astronomy to the Einsteinian revolution in physics. This seems far removed from a philosophy of R&D. Furthermore, the trigger of a Kuhnian revolution in pure science is what he calls a crisis – some anomaly, usually of a theoretical sort, looms so large in the scientific community that it can no longer be ignored – and this sort of intellectual crisis would not seem to be what motivates sudden bursts of creative invention in R&D. Nonetheless, there is an aspect of Kuhn's historicist approach – most explicit and noticeable in the postscript to the second edition of *Structure*[6] and in other writings of that same period[7] – that would suggest that philosophies of pure science and of R&D, respectively, would have to be quite different. What I have in mind is Kuhn's insistence on learning how to do scientific (and technical?) work by apprenticeship in a particular scientific (or technical?) community. Kuhn says: "If this book were being rewritten [rather than just adding a postscript], it would . . . open with a discussion of the community structure of science, a topic . . . [of recent] sociological research . . . that historians of science are also beginning to take seriously."[8] In these communities – and Kuhn says they typically number under a hundred members, often far fewer – students learn by doing problems, and in doing them they are "learning the language of a theory and acquiring the knowledge of nature embedded in that language" at the same time, "by a process like ostension."[9] What I would suggest is that this implies or strongly hints that the learning of language-exemplar-nature fit is likely to be very different in, say, an industrial-science community.

b. Paul Feyerabend's "epistemological anarchism" as espoused in *Against Method* (1975),[10] – his claim that, in science, "anything goes" – might also seem relevant to the no-difference thesis. To those familiar with what goes on in R&D when the push is on to beat the competition – whether an enemy in wartime or just beating a competitor to market – "anything goes" might seem an ideal explanatory device. On the other hand, Feyerabend's intent seems certainly to have been primarily to cast doubt on the purity of so-called pure science. If one were to take seriously this anarchist mode of (non?) explanation of scientific (and technological?) progress, the upshot might be not so much the suggestion of a need for a philosophy of R&D as it would be a reinterpretation of academic science along the more helter-skelter lines of industrial and other "targeted" or "mission-oriented" research. (I find this hint, in both Feyerabend and Kuhn, the most intriguing suggestion to be found in their work, but that is a lead to be followed on another occasion.)

c. If Kuhn's and Feyerabend's claims were not enough to exasperate orthodox positivist philosophers of science,[11] the new breed of sociologists of science have turned out to be even more nettlesome. Here is Rom Harré's summary of the new "anthropological" approach to the study of science:

Laboratories are looked upon with the innocent eye of the traveler in exotic lands, and the societies found in these places are observed with the objective yet compassionate eye of the visitor from a quite other cultural milieu. There are many surprises that await us if we enter a laboratory and study a group of scientists in this frame of mind. The idea that the enterprise can be defined in terms of an idealized epistemology, whether that of experimentally based inductions or of the conjectures and empirical refutations of the logicist philosophers of science, is quickly refuted.[12]

Karin Knorr-Cetina and others, such as Bruno Latour and Steve Woolgar,[13] are saying that, in order to examine science (and technology) in the most objective light possible, we must take up Kuhn's suggestion that scientific communities ought to be the unit of analysis. They then push this suggestion about as far as it will go, namely, to a view of science that – in addition to what Harré says about the view being anti-confirmationist and anti-falsificationist – is radically relativist, relativized to particular communities at particular historical junctures. Some of this "strong program" work in sociology of science has already focused on scientific work in applied settings, and it is easy to see how it could be extended fruitfully into analyses of communities of engineers and scientists in R&D settings. The approach, however, may be too relativistic.

d. Finally, among recent post-positivists, John Ziman has proposed a sociological "radical epistemological relativism" that applies to all knowledge but has special implications for academic science. This, in today's incarnations, Ziman sees as distinct from though related to "big science and technology" especially in R&D settings. According to Ziman,

"Public" knowledge studied in archives such as libraries . . . has a major influence on the "private" knowledge within individual minds. A consensual view is usually introjected into the mental worlds of most scientists in a particular generation, producing the resistance to paradigm change pointed out by Fleck and Kuhn.[14]

Ziman is very clear about the contrast, in his model, between academic science and "big science and technology" as represented best in R&D communities. Of the latter he says:

Science is valued by ordinary citizens, by powerful individuals such as politicians and company directors, and by corporate bodies such as commercial firms and government agencies, primarily for its *use*. It is fostered mainly as a resource to be applied to the furtherance of individual and/or collective activities whose goals are *not specifically the advancement of knowledge*.[15]

This Ziman contrasts with academic science, saying that the popular understanding

. . . takes no account of several attributes of science that often motivate scientists personally, such as . . . the aesthetic satisfaction of "discovering and explaining the marvels and mysteries of the world about us."[16]

It should be noted here that both accounts are products of Ziman's radical social epistemology of science and technology; indeed, it is the *social* dimension that guarantees to academic science the "apparent objectivity" it has in particular scientific communities.[17] I would point out that, in Ziman's terms, this could mean that science and engineering, in particular technical communities, could also have – or aspire to – their own limited objectivity.

2. GOING BEYOND ANTI- AND POST-POSITIVISM: A MEAD-DEWEY APPROACH

I want now to claim that, as helpful as these new approaches are, they do not yet go all the way in showing that the various manifestations of science and technology, from the purest of academic science to the least creative applications of technical knowledge, *must* be interpreted in terms of the particular communities in which they are found. That is, I want to claim that R&D science and technology *must* be understood within the context of R&D communities. I would make the same relativistic claim for academic science in academic communities, but that is not at issue in this paper.

According to George Herbert Mead, it is not just the particular community, as Kuhn would have it, that is the unit of analysis; it is the "social act" (i.e., the *praxis*) of a community that makes it intelligible. According to Mead, "The unit of existence is the act."[18]

Mead illustrates this in his attack on all epistemological accounts of science and its history in "Scientific Method and Individual Thinker" (1917).[19] Mead attacks Kantian transcendentalism, Hegelian idealism, rationalism, neo-realism, positivism, and the empiricism of Hume, Mill, and Bertrand Russell. In most cases he is kind enough to recognize the positive contribution of each epistemological theory, but he maintains that all fail to grasp the primary importance of the social act, of the *praxis* of the scientific community at any particular point in history. For instance, attacking Russell, Mead says:

It would be impossible to make anything in terms of . . . sense-data and of symbolic logic out of any scientific discovery. Research defines its problem by isolating certain facts which appear for the time being not as the sense-data of a solipsistic mind, but as experiences of an individual in a highly organized society.[20]

Again, opposing positivism and something close to behaviorism (or what later came to be called "methodological individualism") Mead says:

For science . . . particular experiences arise within a world which is in its logical structure organized and universal. They arise only through the conflict of the individual's experience with such an accepted structure. For science individual experience *presupposes* the organized structure; hence it cannot provide the material out of which the structure is built up. This is the error of both the positivist and of the psychological philosopher, if scientific procedure gives us in any sense a picture of the situation.[21]

In short, whatever various epistemologies of science can offer, it is never the total picture; they fail to situate their insight within the social act of science as it has been historically practiced – that is, within the *praxis* of an organized group of scientists whose world-taken-for-granted is shaken up by some anomalous and problematic discovery which then sets the community on its way to establish another world, and so on and on.

This view of science – indeed, of creative intelligence in any human sphere – as *group problem solving* Mead shares with John Dewey.[22] However, though the view is universally acknowledged (even praised) when philosophers discuss Dewey or Mead, it is almost never recognized that the view necessarily undermines the individualist epistemology of most philosophies of science – indeed, of most of modern philosophy in general.[23]

Moreover, there is an equally fundamental feature of this approach that even defenders of American Pragmatism often overlook – namely, that the progressive, creative problem solving of a community is, by that very fact, an *ethical* endeavor. Here is Mead on this point:

The order of the universe that we live in *is* the moral order. It has become the moral order by becoming the self-conscious method of the members of a human society. . . . The world that comes to us from the past possesses and controls us. We possess and control the world that we discover and invent. And this is the world of the moral order.[24]

Socially productive, creative intelligence is, by its very nature, moral; impeding social progress is the basic immorality. According to an excellent recent interpretation of Mead (which, on this point, also praises Dewey), "Mead and Dewey developed the premises of their own ethics through criticism of utilitarian and Kantian ethics."[25] Hans

Joas goes on to point out how, according to Mead's account of the social act, Utilitarians, basing their view on self-interest, cannot provide an adequate grounding for the (necessary) altruism of social action, just as Kantians fail to see that the right way to do one's duty is not predetermined but must be worked at in a social dialogue or struggle of competing values.[26]

If we apply this to scientific and to R&D communities (separately, of course), it should be obvious that their callings are noble to the extent that they aim at the creative solution of urgent *social* problems. This, however, is no special reason for pride; all other communities in the larger society have the same noble calling. As one example, artistic communities have often helped those of us in modern societies with more technical, utilitarian goals to see that defining "solving social problems" in a narrowly utilitarian way is itself a social problem.

3. SOME "PROBLEMS THAT BLOCK PROGRESS"

One difficulty with the Mead-Dewey approach is that it is vague. If we generalize from Mead's example of particular scientific communities at particular times to society at large – as both Mead and Dewey repeatedly recommend – it becomes exceedingly difficult to identify which problems block progress. This is an especially urgent problem in modern democratic societies with their myriad competing interests.

Fortunately, I do not have to tackle this large theoretical and practical problem in this paper, devoted as it is merely to pointing out how the problems that block progress in the basic sciences and in the various R&D communities are fundamentally different. It is to the task of identifying some special social problems of R&D communities (or, in one case, of all of modern society as it has come to depend on R&D scientists and engineers) that I now turn.

a. The wise use of R&D. For this I look to the example of genetics in R&D settings. If we include under this heading everything from recombining molecules for agricultural purposes to curing genetic diseases to cloning humans, the problem is sharpened considerably. Some people say that "wise" guidance for industrial scientists and engineers in this area is neither needed nor possible: like any other form of free enterprise, genetic research and development should be subject only to the control of a free market. Others go to the opposite extreme, loudly proclaiming jeremiads against such research:

Ethical sensibility is an essential dimension of being human; it is an integral part of the natural-cultural characterization of humanity. But numerous possible forms of human manipulation run the risk of affecting, altering, even suppressing this ethical capacity. To be convinced of this, it is enough to call to mind manipulations of procreation; increased ability to modify personalities . . . electrochemically; the likelihood of genetic manipulations; prospects of a cybernetic reconstruction of humans in a general fashion – in short, all those manipulations that would affect in profound ways what Jaspers has called "situational limits."[27]

Extreme complaints of this sort against genetic R&D in industry have not moved democratic societies to enforce strict regulations, much less a total moratorium, on for-profit genetic research – though such controls are widely espoused in certain religious circles. What we have seen, in the 1970s, is the mobilization of significant forces attempting to regulate recombinant-DNA research. In that case, the issue was not seen as a threat to humanity but as a question of *democratic control* of what were perceived as potential threats to human health and well-being.

What I see as the potentially stymieing problem facing genetic re-searchers and biomedical engineers in industry is this very diversity of opinion on genetic research. They *must* take into consideration poten-tial opposition to their efforts – especially if these lead to such futuristic possibilities as cloning humans – in ways that scientists doing basic research on genetics in academic settings need not consider. (Admit-tedly, some opponents of genetic research would prefer to stop it in academia as well as in profit-making enterprises, but their voices are not taken very seriously.) A wise manager of genetic R&D would do well to heed the lessons of the recombinant-DNA debate and recognize that directions – possibly even putting a stop to the venture – will inevitably be set in public debate, at least in open, democratic societies.

b. Urgent social problems facing R&D institutions. As an example here I would suggest the demand that corporations provide child care for their employees – a mundane matter, certainly, but also a serious issue of justice for many women. Here again there are a wide variety of opin-ions. What I want to focus on is the potentially serious conflict this issue raises with respect to another demand made of R&D institutions – namely, that they hire more women and minority scientists and engineers.[28] Most people would recognize the latter as a progressive move, but it is hard to see how that goal can be met without some provision for child care for women (perhaps especially minority women) researchers. At the very least, one could say that the opportunity goal would be easier to attain if child care were more readily available.

The sense in which this problem is different for R&D researchers as distinct from academic scientists is that there are so many more opportunities for women and minorities in industry and government R&D. Providing child care for female academic researchers would barely make a dent in the problem; on the other hand, in industry and government, child care facilities would almost certainly be extended to take care of other employees, including men as well as women and minorities.

c. Can R&D be blamed for increasing economic inequities in contemporary society? Albert Borgmann thinks that there is a correlation between technology, as he construes it, and economic inequality. He first notes the commonly recognized wide disparities between the rich and the poor even in the most open of Western democratic societies, where the appeal of egalitarianism is strong and where some would expect that technological growth would have lessened such disparities. Borgmann then asks: "Given that liberal democracy has provided a notion of equality and has advocated it and given that the machinery of government to implement that notion is available, why is inequality so generally accepted?"[29] Many, of course, including many scientists and engineers in R&D communities, would not only accept inequality, but advocate it – as long as it is joined to another notion, equality of opportunity. Borgmann agrees, but gives the answer a different twist:

The peculiar conjunction of technology and inequality that we find in industrially advanced Western democracy results in an equilibrium which can be maintained only as long as technology advances. The less affluent must be able to catch up with the more affluent at least diachronically; they must be able to attain tomorrow what their wealthier neighbors have today. If the general standard of living comes to be arrested for the indefinite future, inequality will become stationary and more manifest and will require a new solution.[30]

The most likely "new solution" would be a revolt of the have-nots against the haves. Instead, as Borgmann sees things, those who have the power fail to act "because inequality is a motor and stabilizer of technological progress, and since the middle class is committed to the latter, it tolerates the former."[31]

This problem for the R&D community is likely to be recognized only by those members who have a deeper commitment to democratic equality than to technological "progress." But all members of technologically advanced societies – and especially the scientists and engineers who are responsible for technological advances – would have something else to consider, not only if the poor were to rebel, but also if technological advance continues, as it seems to be continuing, to increase rather than lessen the differences between the very rich and the very poor.[32]

d. are philosophy and humanistic values generally threatened in technological society? Carl Mitcham reflects the views of many Continental philosophers when he answers in the affirmative:

Two centuries of philosophical criticism [of technology], comparable in character to the centuries of criticism that ushered in the modern period, have been ineffectual in altering the course of modern commitments in any fundamental way. Is this because such criticisms have been misguided? Is it because within the conditions that philosophy helped establish, philosophy has been deprived of its rightful powers? Or have expectations of the powers of philosophy been misconceived?[33]

Mitcham clearly thinks that philosophy ought to retain its culture-critical powers, its power to challenge technology, and his opponents are clearly those analytical philosophers who have abandoned the grand claims of traditional Continental philosophy in favor of an approach similar to that of science: technically precise but not aimed at criticism of society.[34]

How, one might ask, is this relevant to R&D scientists and engineers? One answer lies in historical trends, in the fact that, over the past hundred years, the number of scientists and engineers as a percentage of the U.S. Census Bureau's "professional workers" category has steadily increased while humanists have decreased – and the trend has been more marked since World War II or at least since Sputnik.[35] At the same time, critics have claimed, traditional cultural disciplines have become more and more technical.[36] The problem here might be a return to the first problem above: where will the people come from who will play the role of criticizing goals in a society filled with specialized technical experts?

The way I would relate this problem to the R&D community is in terms of recent cries – most often voiced by spokespersons for that community – for "technological literacy." A need for that there surely is, but members of the R&D community should also recognize the urgent need to educate the young in the humanities and in those civic virtues that John Dewey devoted his life to espousing. Without critics and citizens adept at democratic political give and take, research and development can all too easily lead us up blind alleys such as the nuclear arms race.

CONCLUSION

It may seem obvious (certainly it seems so to me) that scientists and engineers in R&D communities – not only in industry but in hospitals and pharmaceutical companies, in agriculture, in government agencies of all sorts – have particular social responsibilities that are not shared by academic scientists, at least not to the same degree. It may also be the case that the anonymous referee of a book proposal for a larger version of this project who was cited at the beginning was not looking for this sort of thing when he or she asked to be shown something "intellectually challenging" about the difference between academic and R&D science and engineering. Nonetheless, in my Pragmatism-based view, issues of epistemology, of the relationship between theory and fact, and so on, need to be raised *within* the social act context – and that is true for "pure" as well as "applied" research – if those issues are to be debated in a realistic fashion. If the issues are raised in that context, it may not be the case, as Mead and Dewey would have it, that epistemology should give way to social science studies of science and technology. But it is likely to be the case that the resultant epistemology of science or the theory of engineering knowledge would be very different from anything we have seen in recent philosophy of science – even recent philosophy of science of the most anti-positivist sort. In my view, praxical concerns come first, even if epistemological and other analytical issues must be addressed in order to address praxical questions effectively. The priority may be one of place rather than time, but it is there. The proper relationship between facts and values, epistemology and ethics, is dialectical – as a careful study of R&D science and engineering will show.

One final note on the Charles Pedersen Nobel prize example. I think that, despite appearances to the contrary, it does not support Hempel's no-difference thesis. By all accounts, Pedersen was a loyal DuPont chemist committed to the company's slogan, "Better things for better living through chemistry." That his particular situation allowed him the luxury to make a "pure science" discovery in no way seems to invalidate the claim that almost all discoveries and inventions in R&D settings are motivated by, in Ziman's words, "goals other than the advancement of knowledge."

NOTES

[1] See the Wilmington, Delaware, *Evening Journal*, Friday, December 11, 1987, pp. A12–13; also, for some technical detail, *Chemical and Engineering News*, October 19, 1987, pp. 4–5.

[2] See my "Toward a Philosophy of Engineering and Science in R&D Settings," in P. Durbin, ed., *Technology and Responsibility* (Philosophy and Technology, vol. 3; Dordrecht: Reidel, 1987), pp. 309–327.

[3] An anthology of original contributions by authors who have done such work as is currently available should be published in 1989 in the Research in Technology Studies series, Lehigh University Press.

[4] Thomas S. Kuhn, *The Structure of Scientific Revolutions* (2d ed.; Chicago: University of Chicago Press, 1970; original, 1962).

[5] *Ibid*, p. 8.

[6] *Ibid*, pp. 174–210.

[7] See especially Imre Lakatos and Alan Musgrave, eds., *Criticism and the Growth of Knowledge* (Cambridge: Cambridge University Press, 1970), pp. 1–23 and 231–278.

[8] *Structure*, 2d ed., p. 176.

[9] *Criticism and the Growth of Knowledge*, pp. 272–273.

[10] Paul K. Feyerabend, *Against Method* (London: NLB, 1975).

[11] See Israel Scheffler, *Science and Subjectivity* (Indianapolis: Bobbs-Merrill, 1967), and Ernest Nagel, "Philosophical Depreciations of Scientific Method," in his *Teleology Revisited and Other Essays in the Philosophy and History of Science* (New York: Columbia University Press, 1979), pp. 84–94.

[12] Preface to Karin D. Knorr-Cetina, *The Manufacture of Knowledge: An Essay on the Constructivist and Contextual Nature of Science* (Oxford: Pergamon Press, 1981), p. viii.

[13] Bruno Latour and Steve Woolgar, *Laboratory Life: The Social Construction of Scientific Facts* (Beverly Hills, Calif.: Sage, 1979).

[14] John Ziman, *An Introduction to Science Studies: The Philosophical and Social Aspects of Science and Technology* (Cambridge: Cambridge University Press, 1984), p. 110; the second reference is to Ludwig Fleck, *Genesis and Development of a Scientific Fact* (Chicago: University of Chicago Press, 1979; German original, 1935).

[15] *Introduction to Science Studies*, p. 112.

[16] *Ibid*.

[17] *Ibid.*, p. 110.

[18] George Herbert Mead, *The Philosophy of the Act* (Chicago: University of Chicago Press, 1938), p. 65.

[19] Mead, "Scientific Method and Individual Thinker," in G.H. Mead, *Selected Writings*, ed. A. Reck (Indianapolis: Bobbs-Merrill, 1964), pp. 171–211; appeared originally in the collective volume (no editor listed), *Creative Intelligence: Essays in the Pragmatic Attitude* (New York: Holt, 1917), pp. 176–227.

[20] *Ibid.*, p. 196.

[21] *Ibid.*, p. 203.

[22] The best recent account is in R.W. Sleeper, *The Necessity of Pragmatism: John Dewey's Conception of Philosophy* (New Haven: Yale University Press, 1986). Sleeper emphasizes logic, metaphysics, and theory of knowledge, but everything is subordinated to Dewey's "meliorism."

[23] See John Dewey, *The Quest for Certainty* (New York: Minton, Balch, 1929).

[24] Mead, "Scientific Method and the Moral Sciences," in *Selected Writings* (note 19, above), pp. 248–266; the reference is to p. 266. The essay first appeared in the *International Journal of Ethics* 33 (1923): 229–247.

[25] Hans Joas, *G.H. Mead: A Contemporary Re-Examination of His Thought* (Cambridge, Mass.: MIT Press, 1985; German original, 1980), p. 122.

[26] *Ibid.*, pp. 122–124.

[27] Gilbert Hottois, "Technoscience: Nihilistic Power versus a New Ethical Consciousness," in P. Durbin, ed., *Technology and Responsibility* (note 2, above), pp. 69–84; reference is to p. 80.

[28] See, among many other items on this topic, Michael Crow, ed., *Women and Minorities in Science and Engineering* (Washington, D.C.: National Science Foundation, 1977).

[29] Albert Borgmann, "Technology and Democracy," in P. Durbin, ed., *Research in Philosophy and Technology*, vol. 7 (Greenwich, Conn.: JAI Press, 1984), pp. 211–228; reference is to p. 223. Essentially the same material appears in Borgmann's *Technology and the Character of Contemporary Life* (Chicago: University of Chicago Press, 1984), chapter 16.

[30] *Ibid.*, pp. 223–224.

[31] *Ibid.*, p. 223.

[32] Whether the poor are technological illiterates in technological societies or poor countries relative to technically advanced countries.

[33] Carl Mitcham, "Philosophy of Technology," in P. Durbin, ed., *A Guide to the Culture of Science, Technology, and Medicine* (New York: Free Press, 1980; paper, with additional bibliography, 1984), pp. 282–363; reference is to pp. 343–344.

[34] The "technical" credo of analytical philosophy is succinctly stated by Bertrand Russell, *A History of Western Philosophy* (New York: Simon and Schuster, 1945), p. 834, and by Hans Reichenbach *The Rise of Scientific Philosophy* (Berkeley: University of California Press, 1951), p. 123.

[35] Compare, for example, *Statistical Abstract of the United States, 1982–1983* (Washington, D.C.: U.S. Government Printing Office, 1983), with *Historical Statistics of the United States: Colonial Times to 1970*, 2 vols. (Washington, D.C.: U.S. Government Printing Office, 1975); also, Laurence Veysey, "The Plural Organized Worlds of the Humanities," in A. Oleson and J. Voss, eds., *The Organization of Knowledge in Modern America, 1860–1920* (Baltimore: John Hopkins University Press, 1979), pp. 51–106. Veysey carries his contrasts up to the 1960s.

[36] As an old but forceful example, see Jacques Barzun, *Science: The Glorious Entertainment* (New York: Harper & Row, 1964).

PHILIP T. SHEPARD

IMPARTIALITY AND INTERPRETIVE INTERVENTION IN TECHNICAL CONTROVERSY[1]

Impartial treatment of technical controversy confronts a dilemma in the clash between objectivists and normativists. Any given dispute, it might be argued, either involves technical questions or it does not. If it does involve technical questions then the objectivist will have the upper hand. If it does not involve technical questions then the normativist will have the upper hand. Either way the prospects for impartial intervention are either dim or non-existent. This paper aims to develop a way out of this dilemma by exploring a conversational dispute that parallels recent examples from agricultural controversy.

The clash between normativism and objectivism is writ large in recent agricultural controversy in the U.S. Critics of the agricultural research establishment have been preoccupied with the values of conventional agriculture, with diagnosing the ills of conventional practices, and with prescribing alternative values and practices. Almost invariably, they have found the practice of agricultural research ethically wanting and traced its shortcomings to "bad" values. Their prescriptions of "good" values all too often have been placed beyond the reach of factual counterarguments. Questions of economic, social or political feasibility have been given the short shrift, circumstantial constraints ignored, and the complex relations between values and the practices they inform and guide obscured.[2] By seeming to preempt the moral high ground, the critics have maneuvered proponents of the establishment into falling back on appeals to scientific and technical authority. By taking the universal acceptability of their valuations for granted, critics have presumed on the values and character of their opponents, polarized dispute and foreclosed attempts at mediation.

Defenders of the U.S. agricultural research establishment, on the other hand, have regularly castigated their critics as "irrational" and "neo-Luddite" and defended agricultural science by appealing to the authority of scientific method as value neutral and above politics and morality.[3] By taking their own view of science for granted and placing themselves in the position of exclusive arbiters of truth and reason in matters agricultural, agricultural scientists have often dismissed

47

Edmund F. Byrne and Joseph C. Pitt (eds), Technological Transformation: Contextual and Conceptual Implications, 47–65.
© *1989 Kluwer Academic Publishers.*

moral issues out of hand.[4] By promoting particular policies from an ostensibly value neutral stance, they have ignored the relevance of values to decisions made under uncertainty and disdained to confront the non-scientist on equal terms. Thus, defenders of the establishment have also polarized dispute, making impartial mediation much more difficult. While critics have privileged their normative concerns at the expense of factual accuracy, defenders have privileged scientific authority at the expense of moral sensitivity and political accommodation.

To discern how impartial treatment of technical controversy is possible and to reduce disagreement through more effective mediation, we need to understand better how the technical side of controversy interacts with the non-technical side. Impartial treatment must span both normative and factual considerations without presuming on the views of either normativists or objectivists. Appeals to values, or ethical principles, and to scientific authority need to be presented in such a way in the context of controversy that their relevance to the issues at stake can be grounded in interpretations of the issues that are acceptable to all the contestants. Thus are we pressed to find common ground – between scientists and non-scientists, between factual and normative considerations, between opposing viewpoints, and between divergent meanings and conflicting standards of relevance.

Where can common ground be found and how can it be utilized in the interest of mediation? The answer this paper explores is that the common ground lies in the everyday language of appeals to "facts" and "values" and can be brought constructively into play through an interpretive approach to mediation. In sports, in family life, and in business negotiations, for example, we already know to some extent how to mediate conflicting factual and value claims: how they depend on one another, how they speak to one another, how they can be given equal and fair hearings, and how they can lead to compromise and partial resolution if not always to complete satisfaction. Appeals to facts and values have their natural home, not in the discourse of science and technology, but in everyday discourse.

To see more clearly the character and limitations of objectivism and normativism then, and to bring into focus an alternative interpretive style of intervention, an example will be given of interpersonal conflict reflected in conversation. The point of the example is to situate the three strategies of intervention in a context that allows our ear for ordinary language and everyday social sensibilities to function. Ideas suggested by delineating styles of intervention in an everyday context will be tried out in sorting through some of the dynamics of technical disputes.

The conversational example below does differ in some respects from public controversies, especially technical ones. In the public arena, opponents do not often converse with each other directly; more often their differences are expressed in writing, published perhaps in different places. Even when they do come together in one place, disputants are usually put at a social distance from each other by formal rules of court procedure, public hearings, professional conferences, or the like. Certainly these differences have a significant impact on the quality and effectiveness of communication. But they do not reach to the roots of our problem. Even in more favorable conversational contexts, communication breaks down and polarization occurs. So we begin there, in order to find out what the prospects are for impartial intervention.

THE QUARREL

Suppose that Felice and Rollo are on a winter outing in a Northern climate and that Rollo, but not Felice, has never before experienced snow.

Rollo reaches a bare hand into the snow and exclaims, "Oh! The snow is *cold*."

Felice gives Rollo an exasperated look, having argued all morning to get him to risk this adventure, and retorts, "No, it's not. The snow is *nice*."

Rising to the challenge, Rollo says, "Well, it feels cold to me."

"That's all right," replies Felice, "you'll warm up."

Unsatisfied, Rollo asserts, "But the fact is, the snow is *cold*."

Equally stubborn, Felice insists, "No, it's not. You wouldn't know cold snow if you felt it."

This stretch of conversation, this quarrel between Felice and Rollo, could be thought of as a controversy of sorts – one where the number of participants on each side is reduced to a minimum of one each. The conversation can be divided into three moments or stretches: the moment of contestation where Rollo's surprised exclamation is met with Felice's denial and counterassertion, the moment of unfolding the controversy, of objections and answers, and the moment of polarization, of barbed rhetoric and threatened breakdown.

Note that in the conversational example, at the moment of polarization neither Rollo's appeal to "the fact" of the matter nor Felice's attack on his perceptual competence involves any explicit claim to scientific status or expert technical competence. The example underscores the point that the expression "the fact is . . ." has established uses outside of scientific and technical discourse. Appeals to "facts" are not necessarily appeals to science or expertise. Though willingness to accept factual appeals may derive from the authority of scientific institutions, other types of institutional authority may be relevant (the Church, the Government) and other social factors as well such as interpersonal or intercommunity trust.

As in more formal controversies, people with different outlooks may disagree on whether technical issues are involved in Felice and Rollo's quarrel and on what they are. In agricultural controversies, for a parallel example, questions of the quality of life provided by "family farming" are typically treated by alternative agriculturalists as non-technical, moral and political questions, while more conventional agriculturalists usually treat them as technical questions about standard of living and perceived quality of life. Questions of the relevance of the technical are "trans-scientific"[5] and not strictly internal to technical or scientific disciplines. They depend, at least in part, on the social character of controversy.

Moreover, even "technical" facts are social productions.[6] However much one thinks a controversy turns on technical issues, those issues are always located somehow in a social context from which appeals to facts and values take their meaning.

THREE STYLES OF INTERVENTION

The three styles of intervention can be introduced each through a different third person who enters the conversation and acts out the style.

First, consider normativist intervention and imagine the conversation continuing as Felice's acquaintance, Vincent, approaches the pair, having overheard the initial interchange.

Take I: Normativist Intervention

"My dear Felice, how nice to see you on such a balmy day," declares Vincent.

"And you must be Rollo," he says, turning to shake hands with Rollo. "You really must admit this is a great day to be outdoors. Such invigorating weather! So healthy!

"You'll have a fine time, I just know it. Why, no one with any mettle at all could fail to enjoy an outing today."

Vincent's intervention here is friendly enough, but clearly hortatory in character. He is urging a particular view of healthful activity and, in the process, presuming on Rollo and siding with Felice's position. Felice may or may not be happy to have this kind of support: Rollo could conceivably swallow his misgivings and go on to enjoy the outing, but to do so he has now to overcome not only Felice's attack on his perceptual competence but also Vincent's presumptions on his beliefs, values and character. All too often, recent critics of conventional agriculture, like Vincent, have taken their own values for granted and simply prescribed "the good." They have not shied from impugning the character of their opponents.

For the second, objectivist style of intervention, imagine another person, Oliver, approaching Felice and Rollo bearing a thermometer.

Take II: Objectivist Intervention

"I couldn't help noticing," says the stranger, "that you were arguing just now about the temperature of the snow. Since I happened to have a thermometer with me, I thought perhaps I could be of assistance.

"By the way, my name is 'Oliver'."

After putting his thermometer in the snow for a minute, Oliver pulls it out, reads it and declares, "the temperature of the snow is precisely 28 degrees Fahrenheit."

Oliver's intervention is quite civil, and apparently neutral, since he does not advocate or prescribe anything. Oliver refers the issue of coldness back to the "real world" by testing the temperature of the snow with a thermometer. In the present cultural setting, this may appear to be the straight and open path to resolving the dispute. The U.S. Department of Agriculture does something very similar when it refers an issue to a panel of scientists for adjudication.

Oliver's intervention confronts with a significant problem, however, concerning the relevance of his measurement to the issue at hand. Rollo might find it highly relevant: "28 degrees! Why that's below freezing. I was right; the snow is cold." But Felice might object, "We were arguing about whether the snow is *cold* or *nice*, not about what its temperature is!" Instead of resolving the dispute, Oliver's intervention may turn out to be covertly partisan by lending the weight of "scientific" or "technical" analysis to only Rollo's side of the dispute.

For the third style of intervention, imagine the couple approached by a mutual acquaintance, Marian.

Take III: Interpretive Intervention

   "Wait a minute, you two," Marian interjects, "Before you get into a ruckus, maybe you could settle some of your differences if we talked things over a little more.

   "I wonder if you'd mind," Marian says, "if I asked each of you a question?

   "Rollo, you seem to be concerned that the snow is cold. I wonder if what's bothering you is that it might be cold enough to be harmful?

   "And you, Felice, do you feel like you're losing patience with Rollo? I wonder, if you do, if that's because you're pretty convinced that the snow's coldness is not dangerous?"

Marian's intervention acts to forestall polarization and to facilitate dialogue between Felice and Rollo. She creates an opportunity for resolution by her invitation to "talk things over" and by querying the meaning of each participant's position. Her queries may be helpful in two ways: they offer explicit interpretations that can be accepted, rejected or amended as the disputants see fit; and they draw out implicit or covert themes, linking them to the explicit point of disagreement. Rollo emphasized that the snow is "*cold*" and Felice that it is "*nice*." Marian attempts to "unload" their uses of these terms by offering the readings "cold enough to be harmful" and "the coldness is not dangerous." This moves some options closer to the reach of the disputants and gives them the choice to try to understand their differences rather than react to each other from half-conscious feelings.[7]

Some scholarly work on agricultural issues has shared some elements with Marian's interpretive style. For example, in a 1982 conference at the University of Florida, Aiken tries to interpret conflicting positions and provide a "conceptual map" of the different assumptions of disputants; and Hildreth suggests the relevance to policy decisions of both "positive" and "normative" information and a neutral strategy for education on controversial issues.[8] Also, problem solving research sometimes shares elements of this style, for example, the Michigan Pickle Study, which mediated successfully between growers, processors, Mexican Nationals and other interested groups and agencies.[9]

THE STYLES COMPARED

The three interventions here are all well-intentioned, yet differ significantly in their presumptions, their focus or emphasis, and their social consequences. Vincent's normativist intervention presumes on Rollo's values and character. It takes a partisan stance and attempts to snow Rollo with effusive praise for healthful winter weather. Oliver's objectivist intervention tries to avoid such presumption by taking a neutral stance, above or apart from the quarrel, and by appealing to "objective" standards. However, Oliver has presumed that objective measurement of temperature is relevant to the controversy. Marian's interpretive intervention also tries to avoid presumption, but by a different strategy. She invites dialogue and queries for meaning. Like Oliver and Vincent, she also presumes that the couple is willing to open their dispute to outsiders. In agricultural controversy presumptions like Vincent's are likely to be well received only by like-minded advocates; Oliver's will be favored by many agricultural scientists; and Marian's by a variety of participants interested in working out their differences.

In its focus, Vincent's intervention is preoccupied with values, which it assumes as given ("so healthy"), and prescriptions, which it infers without argument ("you must admit"), and tends to ride roughshod over specific countervailing information. Rollo, for example, might have poor circulation or be otherwise unusually vulnerable to frostbite or hypothermia. In agricultural controversy, environmentalists alarmed by soil erosion statistics often condemn continuous corn cropping as one of the culprits without considering information that some methods of continuous corn cropping may actually build up the soil by adding large amounts of plant material.[10]

Oliver avoids Vincent's problems by focusing on a matter of fact and eschewing moral value judgments and prescriptions. Yet his shift takes for granted the epistemic values of evidence and clarity, and the epistemic prescription of testing. If Felice objects that temperature is not at issue, Oliver's epistemic assumptions are likely to become more visible. "My dear woman," he might reply, "if you will kindly just specify what you mean by 'cold', then the truth of the matter can be tested forthwith." Oliver's fastidious civility here masks a semantic prescription, à la logical empiricism, that what meaningful terms ascribe should be testable. Marian's approach, however, invites dialogue, queries for meaning evenhandedly, and explicates links between facts and values ("cold enough to be harmful"). In calling attention to this linkage, it attempts to overcome the one-sidedness of both the normativist's emphasis on values and prescriptions and the objectivist's emphasis on facts and epistemic standards.

In their likely social consequences the three styles are also quite different. Vincent's enthusiasm for winter weather carries an accusatory undertone: it is a deficiency of character not to enjoy such weather. Most people would react to this defensively. Were an objectivist not at hand, Rollo might well draft one for the occasion.

Oliver's "objectivity," on the other hand, ignores, denies or dismisses the social and interpersonal meaning of the controversy. The emphasis, inflection and emotional weight placed on 'cold' and 'nice' (represented by italics) is sliced away by Oliver's prescription of testability, and so are Vincent's values. Most people would bridle at this kind of dismissal; Felice might enlist Vincent's support. In complementary ways, both Oliver's and Vincent's interventions have increased the level of disagreement and fanned the flames of acrimony.

In contrast, Marian opens a path to better understanding. By inviting Felice and Rollo to explore their differences and by suggesting links between implicit evaluations and explicit descriptions (e.g. 'harmful' and 'cold'), Marian gives them a way to understand each other better and to search for common ground. Prospects of resolution at the second moment of the dialogue are brought closer by Marian's intervention. Thus, the chances are increased that Felice and Rollo will resolve their differences, or, at least, reach a more amicable agreement to disagree.

OBJECTIVISM IN SOCIAL PERSPECTIVE

The conversational example provides an opportunity to understand the limitations of objectivism from the perspective of the social dynamics of dispute. Even if we assume with the objectivist that scientific method impartially resolves factual disagreement in science, it does not follow that it will do the same when transposed to other contexts. In spite of its apparent neutrality, Oliver's intervention polarizes the dispute when it lends the weight of "science" to only one side.

What limitations are reflected by the seemingly inadvertent polarization occasioned by even the best-intentioned objectivist intervention? What positive role, if any, remains for the scientist in public controversy? These questions are especially pressing ones for agricultural scientists confronting current political changes. Some answers are suggested by the conversational case above.

Rollo and Felice disagree on whether "the snow is *cold*." Is this a factual disagreement? At the moment of polarization, Rollo asserts that it is. But Felice still disagrees, and she may yet insist that the point of her disagreement is not to dispute a matter of fact at all. In the context of controversy, partisans struggle with each other over what is to be accepted as "fact," as well as over what "values" deserve allegiance and what prescriptions ought to be followed. Only if they perceive some advantage, are they likely, as Rollo does, to accept the objectivist strategy of determining "the facts" without reference to either values or prescriptions.

Quarrels, however, are usually not empty. Setting aside the rhetoric of appeals to fact, we can still ask what is at issue in this quarrel. There are many possibilities, among which are:

a) The temperature of the snow is below freezing.
b) Exposure to the snow is more likely than not to be harmful to humans.

c) Prolonged exposure to the snow will probably harm humans.
d) Exposure of bare human skin to the snow will cause frostbite.
e) Prolonged exposure of bare human skin to the snow will cause frostbite.

Whether one of these, or some other, is a reasonable and acceptable statement of the issue at hand is an interpretive problem. One, moreover, that cannot be resolved impartially without consulting the purposes, concerns, beliefs and values of all the disputants. One could argue that what Rollo meant by *"cold"* is not testable, hence not captured by any of the above readings. Or, one could argue that, say, d) is most plausible in context. But to insist without argument on facile interpretations, like a), risks polarizing the dispute.

Unfortunately, the tendency to ignore differences in outlook and sweep interpretive problems under the rug is not rare among agricultural researchers who have practiced objectivist intervention. For example, Luther Tweeten, Regents Professor of Agricultural Economics at Oklahoma State University, recently addressed in *Science* magazine issues about the "desirable size, number, type, tenure, and legal organization of farms." He focused on a number of "more or less conventional assertions" which he took "as hypotheses to be tested":

1) Small farms provide a higher quality of life to operators and their families than do larger farms.
2) Small farm operators take better care of their soil than do larger farms.
   . . . .
5) Society would be better off if publicly supported research and education were focused on small farms.[11]

In the context of agricultural controversy, statements such as these take on a wide range of meanings depending on the outlooks of the participants. Phrases like 'a higher quality of life', 'better care' and 'better off' explicitly express evaluative judgments; and the values, standards and criteria on the basis of which such judgments are made vary widely among participants. There is considerable difference of opinion, for example, over whether heavy use of artificial nitrogen fertilizers constitutes "good care" of the soil; and these differences arise in part from differences in standards for "good care." By treating such statements as conventional testable hypotheses, Tweeten glossed over differences of standards, values and criteria and assumed or presumed an agreement on a single, testable interpretation of each statement. This is precisely what Oliver would have been doing if he had insisted that the fact at issue in the quarrel was whether the temperature of the snow is below freezing.

Objectivist limitations arise from difficulties in relating purportedly invariant, authoritative, realist and "objective" science to variant, relative, situational and "subjective" aspects of human disputes, particularly the shifting and shifty character of what is at issue. Is the issue factual or normative, or both at once? The disputants may not agree. Is one or another statement of what is at issue the best one? Again, the disputants may not agree. Declarations of "the facts," or of "scientific information," become partial in their social effects when they are imposed on such situations. Contention makes them contentious.

If there is a constructive role for the scientists, other than getting drawn into the fray, it would seem to depend on two conditions: that constraints on disputants' beliefs are adduced without prejudice to any of their positions; and that all the disputants accept the authority of science, or of the particular scientists, in the matter adduced. Only if these conditions are met can the objectivist effect an impartial role; and that role is not to arbitrate, but to help the disputants to locate or relocate the issues and to think clearly about them. [12]

Oliver, for example, might have helped Felice and Rollo to relocate their dispute by pointing out, without prejudicing the issue, that the temperature of snow never actually varies by more than a few degrees. If Felice and Rollo could both have accepted scientific evidence as appropriate for that point, then Oliver might well have succeeded in convincing them their dispute had been misplaced in the issue, simpliciter, whether the snow is cold.

FACT/VALUE INTERDEPENDENCE

The conversational example suggests that objectivists may yet play a constructive role in working through public controversy. But it is a very limited role and it leaves us still looking for an approach to impartial intervention that is free from its limitations and can be used even when disputants contest the authority of science or of particular scientists.

Since normativist approaches are even more limited, it would seem that only the interpretive approach has a good chance to mediate successfully.

Interpretive intervention does not adduce constraints on disputants' beliefs; nor does it appeal to standards that are external to the context of controversy. On both counts, then, it is free from the limitations of objectivism. There is, however, a deeper point that supports and raises hope for an interpretive style. To explain it, let us return to the problem of relocating Felice and Rollo's dispute.

What *is* the import of Rollo's original utterance of '*cold*'? At stake are both descriptive and evaluative components in Rollo's and Felice's conflicting psycho-social sets toward winter weather. For example, Rollo may have been raised by a grandmother obsessed with the idea that exposure to cold causes illness; and Felice may be an accomplished cross-country skier. For Rollo, then, the situation is fraught with danger; for Felice it is ripe with exhilarating prospects. Reformulation of the issue can take some cues from these disparate "meanings" – both descriptive and evaluative – that Felice and Rollo are struggling to project across each other's life worlds.

Appeals to "facts" and "values" are both typical parts of social controversies, whether or not they are also technical controversies. When they talk of "facts," disputants are engaged in promoting their perceptions to the status of making public claims for each other's acceptance; similarly, talk of "values" involves promoting valuations to the status where they make public claims for allegiance. Moreover, these two kinds of appeal are typically interrelated and interdependent in the dynamics of controversy.

To claim something as fact, or as value, places it in the framework of a particular language game by invoking relations between what is claimed and a language community that includes oneself. When Rollo claims it is a fact that the snow is *cold*, he is appealing to the shared beliefs and abilities of a language community, wider than himself and Felice, and asserting that they would judge him correct on the point at issue (whatever it is) if they had the same opportunities to perceive and evaluate the situation that he does.

Rollo's emphasis on the term '*cold*' and his intonation and inflection (represented by italics) reveal or signal that his evaluations have influenced his interpretation of his experience of the snow. He thinks that others will share his evaluations and corroborate his perceptual judgment, much as we would have done if he had stuck his hand in a fire and said, "It's *hot!*" This comparison reveals something of the extent to which claims about fact depend for their acceptability on judgments of value that are implicated by the context. "Fire is *hot*" is an acceptable factual claim partly because of the shared belief that when human flesh is exposed to fire it gets burned, and partly because of the shared valuation that getting burned is bad.

Felice's part in the dispute illustrates an opposite kind of influence whereby claims of fact can affect claims of value as well. Felice asserted that the snow is *nice*. This statement expresses an evaluation; and one, moreover, that would appear to have been affected by the facts of her skiing experiences – those, especially, that are accepted by other skiers. That she has often skied bare-handed in similar weather without getting frostbitten – a fact accepted by other skiers – encourages her to evaluate the snow as *nice* and supports her in putting this evaluation forward publicly as worthy of acceptance by others. If her experiences were not consistent with this disavowal of danger, then Felice would not be likely to find the snow nice or to urge her assessment on Rollo. So the success of Felice's claim that the snow is nice depends in part on what factual claims implicated by the context are acceptable.

In the context of the language game of making public claims, and hence in the context of public disputes, what is credible in the way of facts depends on what is credible in the way of values, and vice versa. Claims to fact or value are linked, in the pragmatic dimension of their meaning as performatives,[13] to at least some of their opposite numbers. They depend on each other for their warrant as public claims, because they participate *together* in the game of public claim-making, *or not at all*. Moreover, the interdependence of facts and values is not *a priori*, but depends on what we know how to do with our native language. Asserting our perceptions as facts, or our valuations as values, is a social act and what it means depends on what we, the members of a certain language community, know how to do in the frame or frames in which we can interpret it as a meaningful social action – in our repertoire of language games. The interdependence of facts and values, thus, is a matter of practical knowledge which we learn by participating in a language community.

Because it avoids presumption and focuses on clarifying disagreement by drawing out implicit links between facts and values, the interpretive approach is well positioned to engage participants in working through their differences. Instead of adducing constraints on belief and appealing to standards of objectivity that may be foreign or technical to some or all disputants, interpretive intervention relies on what native speakers already know how to do with their language and appeals to their perceptions of fairness, not some "reality" that one must be privileged to speak about authoritatively.

Interpretive intervention also affords the flexibility to respond to different moral viewpoints, outlooks or frameworks. By helping participants to feel heard and to hear each other, it can raise the hope of reconciliation and activate participants' constructive motivations. Instead of reinforcing the impulse to mobilize for political conquest,[14] the interpretive approach teaches us how to live with our differences, and how to search for common ground and a way forward using the medium of a "common sense" embedded in everyday speech.

In short, the promise of interpretive intervention lies in its ability to work with the interdependence of facts and values in the context of making public claims. It frees us from the impulse to privilege facts over values, or values over facts, and offers to all sides in dispute the chance to hear and address each other more effectively. It teaches us how not to presume, by identifying what questions not to beg (because the answers are disputed) and what not to take for granted (because others will not grant it).

IMPARTIALITY RECONSIDERED

If interpretive intervention offers a way to get past the clash of normativism and objectivism, it is in large measure because of the conceptual difference between impartiality and objectivity. Impartial intervention in controversy is a situated achievement that is largely independent of objective knowing. This is reflected in four criteria of impartiality:

1. Participants *perceive* themselves to have been dealt with impartially.
2. The treatment of each side is symmetrical or evenhanded, that is, whatever "moves" are made in dealing with one side are fair game for dealing with any other side – "turn about is fair play."
3. Participants feel they have been understood – that not only have their recommendations been taken into account, but that their concerns and their recommendations have been respected and, in part, incorporated into the outcome.

4. Resolution is not imposed from outside – often, perhaps usually, participants in successful mediation do not perceive the means of resolution as having been imported from outside of the context of the controversy and imposed on it or as reaching beyond their ken, beyond what they already know, to some extent, how to do.[15]

Each of these conditions reveals that achieving impartiality is a matter of the perceptions of participants and is relative to the situation. While objective knowledge may help, it is neither necessary nor sufficient to that end.

While recognizing the contextual and situated character of impartial treatment can free the hand of the mediator, it also raises objections. Believing that only the technical aspects of controversy afford the chance of rational intervention, objectivists will press the mediator to interpret the controversy in a way that maximizes the technical aspects. In effect, objectivists will say to the mediator, "Be more objective." Normativists, on the other hand, will resist the intrusion of the technical in controversy. In the "technicalization" of issues, they will argue, moral arguments masquerade as technical ones so that the masters of technical discourse, or their masters in turn, can prevail. In effect, normativists will say to the mediator, "Be more normative."

Both normativists and objectivists find no middle ground between the position of value neutrality and the position of advocacy. Objectivists will argue that if the mediator's position is value laden, then it must be partial; normativists that mediators appear not to advocate only because they have not owned their own values. Both miss the point.

Impartiality is independent of the objectivists more global or absolute sense of value neutrality. The claim of impartial intervention is not that mediators' statements or methods are free from moral value judgments, but that the outcome of mediation has constructively affected disputants. To check a claim of impartial treatment, one must examine the outcome of the intervention to see what effect it has had and how it has been received by the participants. While the mediator may find it necessary as a practical matter to suspend judgment on the particular claims of both value *and* fact that are implicated in the dispute, she or he may find it equally necessary to underscore judgments of fact *or* value that participants happen to share in common. Whether to suspend or reinforce judgment, whether to reassert or to withhold affirmation, depends on the outlooks of those who participate in controversy, not the analytic character of the issues as disembodied statements.

The mediator's role is to interpret and describe participants' positions to and for each other, perhaps suggesting avenues of reconciliation. When pressed to own her own values, the mediator can do so without sliding into advocacy because her values enter the process of mediation only as heuristic aids in the search for mutually intelligible meanings. To mediate is not the same as to adjudicate the normative issues in the controversy. Interpretive interventions that reduce the level of disagreement and increase the prospects for resolution do so by effecting better communication, not by divesting participants of their positions as judge or jury.

The introduction of technical arguments need not prejudice the issues at hand. If the matter of relevance is addressed forthrightly in everyday language, then questions of the relevance of technical arguments can be adjudicated on their merits by non-specialists as well as by specialists. Technical arguments may or may not distort communication, but whether they do depends on the context and situation – on their positioning with respect to the interests that lie behind discourse and the actions that discourse rationalizes before or after the fact, as well as with respect to claims of fact and value. Interpretive interventions can, and may need to, make the setting of controversy explicit along with participants' outlooks, but mediation is not thereby committed to a particular normative outlook or to a reductive determinism in which the real action takes place "behind the backs" of the participants in controversy.[16]

Interpretive intervention is a workable approach to mediation because it operates on the basis of assumptions that are less loaded and more evenhanded than those of objectivists and normativists. Through a situational approach, it frees the mediator from the pretense of formulas to fit all situations. By centering on the perceptions and outlooks of participants, and on avenues to better mutual understanding, it avoids trying either to settle factual disagreements or to establish philosophical foundations.

ROOM TO MOVE

The reflections offered in this paper certainly do not remove the tensions between normativist and objectivist approaches to dispute. But perhaps they do offer us a way past the dilemma we started out with and room to move in adopting a mediatory approach to controversy.

Technical aspects of controversy, themselves often of disputed relevance, need not present a wall between lay moral judgment and expert opinion, or between activist fervor and practical caution. In public controversy, facts and values are equally common coin, and equally disputed. In them we may find a *modus operandi* for managing disputation constructively, if we can recognize and work with their interdependence as public claims for acceptance and allegiance.

It is, perhaps, not too much to ask of the scientist in the public arena that he or she explain technical concerns and support claims of relevance in the language of everyday discourse. Perhaps it is also not too much to ask of the advocate that he or she address arguments for favored prescriptions in language that speaks to other moral viewpoints and acknowledges shared circumstance. These burdens press neither the objectivist nor the normativist to give up their philosophic orientation or to concede the political contest to their opponents. Both responsibilities, indeed, are already implicitly accepted and invoked by the act of making a public claim.

Where technical questions are relevant, the objectivist has no upper hand because relevance must be established outside the circle of technical explanations in the common understanding. Where technical questions are not relevant, the normativist has no upper hand because alternative values and favored prescriptions must survive the scrutiny of different moral viewpoints and the test of practicality in shared circumstances. The prospects of impartial treatment and eventual resolution or compromise thus remain open whether or not technical questions should prove to be relevant in dispute.

The fact that people engage one another in controversy, including technical controversies, suggests that they have something in common, some common ground from which to address and attempt to work out differences. Common speech reflects this through the interdependence of factual and normative claims. In what we already know how to do with our native language is vested a kind of social wisdom that may well outrun what any one person can effectively express. But if we step aside from the fray, to search for mutual understanding of differences, perhaps we can recover some of that wisdom and find room to move toward reducing disagreement and solving shared problems.

NOTES

[1] This material is based upon work supported by the National Science Foundation under Grant No. RII–8409919. Any opinions, findings and conclusions or recommendations expressed in this paper are those of the author and do not necessarily reflect the views of the National Science Foundation.

[2] See, e.g, Conviser and Freudenberger in Richard Haynes and Ray Lanier, eds. *Agriculture, Change and Human Values*: *Proceedings of a Multidisciplinary Conference*. 2 Vols. Humanities and Agriculture Program, University of Florida, 240 Arts and Sciences Bldg., Gainesville, FL 32611, 1982.

[3] R.C. Rautenstraus, "Public Responsibility of an Agronomist – a University President's View," in *Agronomy*: *Solving Problems, Serving People*. ASA Special Pub. #37. Madison: Amer. Soc. for Agronomy, pp. 1–6, 1980; Robert H. White-Stevens, "Letter to R.P. Upchurch," in C.A. Black, "Informing the Nonagricultural Public about Agricultural Science," *Special Publication* No. 2, (December), 1972, pp. 3–7.

[4] Frank H. Baker, "CAST and the Big T: Truth," Council for Agricultural Science and Technology, Paper No. 7, (November 19), 1979; Theodore Hutchcroft, "Responding to Media Cheap Shots: Observations on the CAST Experience," Council for Agricultural Science and Technology, paper No. 16, (February), 1983.

[5] A. Weinberg, "Science and Trans-science," *Minerva* 10, pp. 209–222, 1972.

[6] Bruno Latour and Steve Woolgar, *Laboratory Life*: *The Social Construction of Scientific Fact*, Beverly Hills: Sage, 1981; Karen D. Knorr-Cetina, *The Manufacture of Knowledge*: *An Essay on the Constructivist and Contextual Nature of Science*, New York: Pergamon, 1981.

[7] The conception of interpretive intervention represented here draws on both humanistic psychology (Carl Rogers, *On Personal Power: Inner Strength and Its Revolutionary Impact*, New York: Delacorte, 1977; Abraham Maslow, *Toward a Psychology of Being*, Princeton, NJ: D. Van Nostrand, 1968) and interpretive social science (Paul Rabinow and William M. Sullivan, *Interpretive Social Science: A Reader*, Berkeley: U. of CA Press, 1979).

[8] Haynes and Lanier, *op. cit.*, pp. 30, 60.

[9] Glenn L. Johnson, *Research Methodology for Economists: Philosophy and Practice*, New York: MacMillan, 1986, pp. 161–169.

[10] The example is drawn from conversation with Glenn L. Johnson.

[11] Luther Tweeten, "The Economics of Small Farms," *Science* 219 (March 4), pp. 1037–1041, 1983.

[12] C. Weiss, "Research for Policy's Sake: The Enlightenment Function of Social Research," *Policy Analysis* 3, pp. 531–545, 1977.

[13] On the concept of "performative" utterances, see William P. Alston, *Philosophy of Language*, Englewood Cliffs, NJ: Prentice-Hall, 1964, pp. 34–36; John L. Austin, *How To Do Things With Words: The William James Lectures delivered at Harvard Univerisity in 1955*, NY: Oxford University Press, 1965; and John R. Searle, *Speech Acts*, Cambridge, England: Cambridge University Press, 1970.

[14] See Philip T. Shepard, "Moral Conflict in Agriculture: Conquest or Moral Coevolution?" *Agriculture and Human Values* 1 (Fall, 1984) :17–25.

[15] Philip T. Shepard and Christopher Hamlin, "How Not to Presume: Toward a Descrip-

tive Theory of Ideology in Science and Technology Controversy," In *Science, Technology and Human Values*, Volume 12, Issue 2 (Spring 1987), pp. 19–28.

[16] Clifford Geertz, "Ideology as a Cultural System," in David E. Apter (ed.), *Ideology and Discontent*. New York, Free Press, 1964; Anthony Giddens, *Central Problems in Social Theory: Action, Structure and Contradiction in Social Analysis*, London: Macmillan, 1979.

J.A. CRANE

THE PROBLEM OF VALUATION IN RISK-COST-BENEFIT
ASSESSMENT OF PUBLIC POLICIES

1 PRELIMINARIES

The valuation components of risk-cost-benefit analysis (RCBA) and related forms such as quantitative social program evaluation involve the selection of the assessment criteria to be employed, the determination of the sizes of benefits and unfavourable outcomes that are to be taken as significant (I'll refer to these as "effect sizes"), and the specification of the probabilities of errors of several kinds that are to be tolerated.[1] "Measurement", on the other hand, might be defined as the specification of the variables required to represent the assessment criteria, and the construction of quantitative indices by which to estimate the values of these variables. Valuation and measurement are thus closely intertwined; neither can proceed without the other, and often they are difficult to distinguish. Yet researchers cultivate measurement and shun valuation. Evidence for this assertion, taken from a recent review of some 40 risk assessment projects:[2]

1. In 19% of the studies, no explicit valuation procedures could be identified; in 51% of the studies, the use of valuation was debatable or ambiguous, since it was limited to predicting effects on prices, profits or other economic variables; in 22% of the studies, valuation was limited to comparisons of risks with other known risks, or to statistical conventions; in only 9% of the studies was there any specific reference to any of the value judgements mentioned above.
2. Account was taken explicitly of any of the following valuation criteria in a fewer than 10% of these projects: provision for principles of social justice in policy – informed consent of risk bearers, compensation, consideration of ability to bear risks – effective involvement of risk bearers in policy formulation, inclusiveness of risk protection over populations at risk, fairness of outcomes of policies. Not only were these principles seldom employed, they were rarely referred to, and there was almost no discussion of value principles in general or of the problem of selecting assessment criteria. This was rarely dealt

67

Edmund F. Byrne and Joseph C. Pitt (eds), Technological Transformation: Contextual and Conceptual Implications, 67–79.
© *1989 Kluwer Academic Publishers.*

with as in any way problematic, aside from practical difficulties of measurement.
3. Though recognized theories and standards of measurement were frequently appealed to, no such theories or standards of valuation were mentioned. It was evident that the studies were guided by no generally accepted methodologies of valuation.

Thus there is good reason to believe that in research concerned with risk, cost and benefits, valuation as defined above plays a surprisingly minor role. Since the selection of assessment criteria and the setting of effect sizes, as well as Type 1 and Type 2 error probabilities are part of the quantitative training of researchers and are stressed as important in the methodological literature[3] it seems unlikely that researchers view them as unimportant. Why, then do they largely disregard them? And what may be done about it?

Concerning these problems, this paper attempts to make the following points:

1. From evaluation as it is practiced one could infer that news of the demise of the fact-value dichotomy, that is reported to have occurred after the mid-1930's,[4] has not yet reached the evaluation research community. That is, researchers still cling to the view that once a research problem has been selected, and assessment criteria decided upon, a project can be designed and carried out with no further reference to valuation.
2. However resistant to change this assumption is, it may possibly be modified by analyses of practice that show how the valuations against which the researchers try to arm themselves steal back in under the radar. Worse, they can distort the researchers' findings, to say nothing of reducing the researchers to the undignified role of dis- coverers of their own assumptions.
3. Researchers might find it easier to pay attention to valuation prob- lems if they could be shown methods for dealing with them that are similar to the methods they are already familiar with and think of as appropriate for themselves given their self-identities as scientists. A potentially useful method for this purpose is magnitude scaling, a method that has already proven useful in the scaling of human valuations of social issues.[5] From examples presented below it is easy to see that the use of magnitude scaling to develop value scales of acceptability of errors, effect sizes, etc. is methodologically similar to

researchers' attempts to measure other theoretical constructs. Therefore, valuation need involve no radical departure from what researchers are already doing and no threat to their self identities.
Many existing social science "measures", indeed, might better be
characterized as exercises in valuation.[6]
4. The proposed method should make it easier to give adequate weights
   to non-market variables with no necessity to monetize all variables.
   An improved aggregation rule is proposed.

To understand the seeming paradox of evaluation researchers who
avoid valuation, I turn to some of their assumptions about it.

2 RESEARCHERS' ASSUMPTIONS ABOUT VALUATION

2.1 Evaluation researchers as disinterested scientists

As noted, reports of RCBA projects seldom contain discussions of
philosophy of valuation. This reflects time and budget restrictions, but
more important, it probably reflects researchers' tacit shared under-
standings that seldom need to be put into words. Methodological
writings are more helpful. A paper by Gibbs entitled "Evaluation
Researcher: Scientist or Advocate?"[7] is a useful example. Gibbs presents a point of view concerning valuation that seems closely aligned
with the practices summarized above.

He identifies himself with the view that researchers should be "disin-
terested scientists". Since evaluative research is the application of
scientific methods, he reasons, the roots of evaluation must be deep in
the philosophy and method of science. "Objectivity", being one of the
basic tenets of science, must also be a cornerstone for evaluation. To
support this contention, he attempts to refute several arguments to the
effect that objectivity in evaluation is impossible to attain.

2.2 "Evaluation" vs "Valuation"

The first of these arguments is that the word "evaluation" entails an
inherent bias because it implies "valuation". The former term means, to
Gibbs, "the resolution of a question of fact" and the latter term means
"the resolution of questions that are based upon an underlying premise
or predisposition about the world that cannot be reduced to a test".[8]
Examples of evaluation questions, according to Gibbs, are: whether
Thorazine has effects on the rate of hallucinations of schizophrenics; the

effects on the frequency of rape of pressuring judges to give lengthy sentences for this offense. Examples of valuation questions: whether schizophrenics should be allowed to hallucinate, and whether the problem of rape in a given county is important enough to warrant expenditures to reduce its frequency.

Gibbs holds that the argument that because evaluation research depends on valuation questions it cannot be objective is fallacious. The fallacy arises from expecting the evaluator to resolve such questions. Gibbs avers that:

If researchers were morally superior to other citizens, they would justifiably be listened to in preference to other citizens regarding valuation questions. The question of who should select what outcome criteria is beyond the special expertise of the evaluator to decide because there is no test, no operation, and no procedure for which researchers have been specially trained that would prove one outcome criterion to be more right or fair than another. Thus questions of valuation might better be left up to politicians, administrators, clergymen and interest groups whose roles do not include any special claim to objectivity.

Therefore, researchers should stick to questions of fact.

2.3 Conservative bias?

Gibbs finds that the argument that because evaluations have an inherent conservative bias they cannot be objective is likewise refutable. This argument hinges on three claims: that in practice researchers are most often commissioned to test traditional programs that are conservative in nature; that merely undertaking evaluation tends to insulate programs from criticism; and that outcome measures are designed by "managers of traditional programs who don't want change".[9] Gibbs counters that since evaluations typically show negative results they can hardly be biased in favour of existing programs. More important, if researchers strive for the same degree of rigour no matter who posed the evaluation question, they will avoid bias. By being careful to limit their inferences to "evaluation" issues, and avoiding "valuation" issues, they will also avoid bias. Thus, they should report differences in results among samples but should not make statements such as "this program is worthy of continued support".[10] By these means objectivity, according to Gibbs, may be preserved.[11]

2.4 Gibbs' assumptions

Clearly, Gibbs believes that a sharp distinction between questions of fact and of value can be made, that the former class of questions can be answered "objectively" and that the latter cannot, and that evaluation research must be concerned only with the former. Once an evaluation problem has been selected, and assessment criteria defined (not by the researcher), no further valuation issues arise.

This helps to explain why researchers try to avoid valuation. I now attempt to show that these efforts not only fail but sabotage the evaluations.

3 DISGUISED FORMS OF VALUATION

3.1 "Facts" vs. "Values"

The claimed distinction between facts and values needs no detailed treatment here[12] beyond noting that while it presents no difficulties to Gibbs, it has bothered many other people, and has proven in practice to be very difficult to maintain. Good evidence of this is given by Mackenzies' case study of the failure of the program to develop behaviourist psychology by strict adherence to the principle that only directly observable behaviours could be admitted to the theory.[13] Mackenzie notes that where this effort was successful it lead to an aimless collection of findings[14] and efforts of the neobehaviourists to correct this came to grief from their inability to find a reliable criterion for distinguishing questions of fact from questions of theory or value.[15] Mackenzies' more general conclusion is also of interest:

"(Behaviourism) is important (to psychology) because it was the most sustained effort ever made to construct a science of psychology through the use of detailed and explicit rules of procedure, because these rules were the outcome of the most sophisticated and rigorous analyses of the logic of science ever made, and because both the attempt and the movement were ultimately failures. The failure was total, even though behaviourism seemed to have every possible advantage."

Of more direct concern here than the general problem of the fact-value dichotomy is Gibb's claim that research design involves no valuative judgements. I have attempted to show elsewhere how design decisions and valuation are intertwined.[16] Here I attempt to show some of the guises under which valuations reappear when researchers attempt to exclude them.

3.2 Valuation by doublespeak

This term is taken from a U.S. Court of Appeals judgment disallowing the New York State's Westway project, that involved the construction of a causeway on the Hudson river.[17] Part of the evidence before the court was a set of regulations by the Army Corps of Engineers barring dredging operations when these would cause or contribute to "significant degradation" of the waters.[18] These included "significantly adverse" effects on fish populations. Another part of the evidence was a later draft environmental impact statement by the Corps predicting that the Westway project would have "significantly adverse impact . . . though not critical". In a supplemental impact statement, the Corps predicted that the impact would be "perceptible", but "difficult to discern from normal yearly fluctuations", and would be "insufficient to significantly impact" the commercial fishery. Asked by the Court to explain the difference between the two statements, the Corps responded that it had used the word "significant" in its technical, statistical sense in the first report, and had employed this term in its common meaning in the final report. Describing this as "Orwellian-like doublespeak", the Court rejected Westway.[19]

Valuation by doublespeak is in use whenever a finding is declared to be "statistically significant" with no discussion of the levels of error appropriate for the problem, and no indication of the effect sizes required for statistical significance, and how these compare with the effect sizes required for policy significance. When results are reported in this way an impression is left that a finding of some importance has been made, when this may or may not be the case. Challenge a statistician on this practice and you will likely be told that it reflects the carelessness or incompetence of some researchers. But in a survey conducted in 1982 of a sample of 1025 instances of statistical inference published over a period of 40 years by researchers in many different fields, all 1025 were found to be of this kind.[20] The 100% prevalence of this practice in the sample suggests that it is what researchers and editors regard as normal practice.

3.3 Valuation by accident

If valuations are not made deliberately, they become accidental by-products of other factors, particularly sample sizes, measurement precision and Type 1 and Type 2 errors.[21] If not planned with valuation in mind, these factors are determined by available resources, expediency, convenience, or convention. The real valuation decisions in these circumstances are hidden; likely, they are not at all what the researchers have in mind. Thus in a statistical power analysis of 25 major findings of social work experiments it was found that the sample sizes and measurement precision were such as to make most of the studies sensitive only to very large differences, for example differences of the order of 40% among samples.[22] In social experiments, much smaller differences are important; differences of 40% are a black hole. Yet the research reports made no mention of the insensitivity of their research to anything less than very large effects. This seems to be an unintended consequence of paying little attention to the problem of valuation at the design stage.

3.4 Valuation by invisible standards

If the valuations of needed effect sizes are often arrived at by accident, valuations of errors are often determined by standards that are unrecognized or invisible to the researchers themselves. Thus, in the analysis of findings of social experiments, previously cited[23] a calculation was carried out of the ratios of Type 1 error probabilities[24] to Type 2 errors.[25] It was found that the researchers were implicitly valuing protection against errors of Type 1 at from 8 to 15 times the value placed on protection against Type 2 errors. Again, this was not part of the research record – it had to be wrung out of the data by secondary analysis.

Such standards seem to be unintended results of the traditional norms of "pure science" on which the researchers attempt to base their work. By the "pure science ideal"[26] it makes sense to guard against Type 1 errors first and give Type 2 errors several decades in which to correct themselves; this protects against cluttering up science with false positives. But this strategy is seldom appropriate for policy studies. To underestimate a risk, for example, may in many circumstances have worse consequences than to overestimate it; to fail to identify positive effects of a social program may often to a worse error than to identify benefits that prove to be illusory. Gibbs is correct in pointing that in some senses evaluation research cannot be accused of a conservative bias, but perhaps a more important sense in which this research tends to be unwittingly conservative is in its use of these pure science conventions that bias against discovering risks of current policies and the positive outcomes of social innovations.

3.5 Valuation by unidentified persons

The failure to recognize valuative judgements leads researchers to reduce them to problems of technique, that by definition are dealt with by technicians or the researchers themselves. The issue of choosing the proper persons to make these decision never arises. The consequence of this is ironic, for those who insist on preserving the scientific neutrality of the scientists, since it leads the latter to an unthinking arrogation of responsibility for moral judgements that should be made by tribunals composed of both scientists and citizens.[27]

As it offers a way of bringing valuation decisions into the light, so that they may begin to be made by explicit rules, I turn now to a look at the possible contributions of magnitude scaling.

4 MAGNITUDE SCALING: A WAY OUT OF THE VALUE-FREE BOX

4.1 Origins
The beginnings of magnitude scaling as applied to social judgements were in the experiments of S.S. Stevens on subjective assessments of sensory stimuli of varying magnitudes.[28] A few of the results of this work were:

1. over some 14 psychophysical continua, including loudness, pitch, weight, warmth and pressure, S.S. Stevens and associates showed experimentally that the strength of a person's subjective sensation of these stimuli is a power function of the strength of the stimulus as independently measured. That is, the data showed an excellent fit to a model of the form:

   $I = c_1 * S^n$

   in which:
   I is the judges rating of subjective intensity of the stimulus and S is the independently measured strength of the stimulus. This finding was later extended to a list of 32 such stimuli.[29]
2. Subsequently, a number of social science researchers applied Stevens methods to judgements of social variables such as a person's income and education.[30] One of the early findings, which has stood up well, was that a power function describes judgments of social status made on the basis of data showing varying levels of income and education.[31] A power function was later shown to hold for a large number of social variables in various domains, including poverty, aggression, income status, occupational status, economic power, merit of teaching, and political support.[32]

3. In all such experiments, samples of judges made ratings in numerical form, drew lines, or used other quantitative means such as pressure exerted on a hand dynamometer to express their subjective impressions of the strength of a series of social variables, expressed in quantitative form, such as income levels.[33]

4. In response to early criticisms that what was being measured was little more than the judges' familiarity with numbers and calculation, it was shown that the same findings are obtained when, when the subjective ratings are made by drawing lines, adjusting sound or pitch levels or pressure.[34] Furthermore, the results were shown to be largely independent of the device used.[35]

5. In the last 25 years many similar applications of psychophysical scaling, other than magnitude scaling have evolved. Now a variety of scaling methods and rationales are available, and much has been learned about the relationships between different forms.[36]

4.2 Example: valuation of poverty from data on income

As noted, much of what is called "measurement" in the social sciences might better be called "valuation". A good example is "measurement" of poverty. Clearly, the variable being measured in this case is heavily value-laden. Using magnitude estimation, Rainwater obtained estimates of "poorness" based on family income from a probability sample of adult housewives in the greater Boston area.[37] He plotted the estimates of poverty against the income data used as stimuli. The plot showed a close fit to a power function: the regression line drawn on logarithmic coordinates through the two sets of data showed a linear correlation of .99.[38] The plot showed that for Rainwater's respondents, poverty or "poorness" begins at median family income levels for the Boston area, strongly supporting Rainwater's theory of the relativity of poverty attitudes, and also providing a useful estimate of the zero point on the poverty scale.

4.3 Applications to the problem of valuation

A straightforward application of magnitude scaling to valuation would be to use assessment variables, such as change in family income, as stimuli presented to a sample of judges on which they would make valuations reflecting the assessment criteria, such the effectiveness of a social program in reducing poverty. This would be a way of working researchers out of the value-free box by means that are congenial to their ways of thinking.

Some of the benefits of this methodology may be:

1. a reasonable measure of "minimum required effect size". To show this, imagine a program to raise the income of welfare recipients above the poverty level and to lessen their dependence on welfare by providing job training. Define the "required minimum effect size" of a proposed increase in welfare clients income as the increase at which the corresponding value scale, i.e. poverty, is zero. It is reasonable to take this increase in income as the criterion for the required "minimally adequate" or "threshold" size of benefit of the program to be evaluated.
2. This could be seen as one of the criteria of the effectiveness of the job training and increased benefits experiment. "Effectiveness" is however not the only assessment criterion that should be employed. Using the same scaling procedures, valuation scales for the inclusiveness, adequacy, equitableness and democratic responsiveness of the program, as well as risks, could be constructed using the same principles of psychophysical measurement.
3. General use of the proposed method would lead to the elimination of valuation by accident and the various other distortions that appear to be widespread at present.
4. The method should also make possible the use of an improved decision rule in risk-cost benefit assessment. This is further pursued below.

4.4 Unexplored areas

Before the proposed method can be fully evaluated, experimentation is needed to determine the full range of assessment criteria to which it can be successfully applied and to test alternative methods of selecting the sample of judges. A working assumption, consistent with much of the attitude scaling research in which the magnitude scaling has been employed, is that the judges must have a first hand familiarity with the variable being scaled. I can find little evidence that the effects of this variable have been systematically explored. Similarly, experimentation on the effects of different modes of presentation of stimuli and of judges' characteristics, including convictions about and active prior involvement with the problem being evaluated, need further research.

Moreover, there are at this stage a number of other unknowns about the effects of scaling procedures. These include the effects of the number of stimuli (e.g. range and number of income levels) presented to the judges and the use of different kinds of scales (e.g. adjective scales) as stimuli. But the substantial body of research reporting successful applications of the method in valuation categories such as poverty, desirability of occupations, power, and political performance[39] suggest that such enquiries would be worthwhile.

5 A POSSIBLE DECISION RULE FOR RISK-COST BENEFIT ANALYSIS

A standard text on Benefit-Cost Analysis points out that monetary valuation of social impacts is rarely possible.[40] As a result of this limitation of method, Thompson recommends, as do most standard texts,[41] that instead of being included in the benefit-cost calculations, distributional effects, especially social impacts, should be presented as descriptive information on the individuals who bear costs and realize gains. Shrader-Frechette, however, has made a powerful case[42] that this kind of compromise undervalues social impacts and therefore seriously weakens if it does not invalidate the conclusions drawn. Quantifying social impacts would have reversed the conclusions reached in a representative group of major technology assessment studies reviewed by Shrader-Frechette.[43]

Thus there is good reason to believe that normal science RCBA leads to a neglect, distortion, and/or systematic underestimation of social impacts. In many cases this amounts to underestimating social risks, and consequently overestimating net benefits. The introduction of the proposed valuation procedures applicable to a large number of social variables permits quantification of these variables, and at the same time suggest a modified decision rule for RCBA, based on complementary quantitative valuation of social and money variables. The procedure would be as follows: first ensure, through the use of the value scales, that effect size and probabilistic error specifications regarding social benefits and risks are satisfied; from policy options that meet this requirement choose the option with the best benefit/cost ratio on money variables. There would be no need to monetize all variables; these would come into play after quantitative social assessment criteria were satisfied.

APPENDIX A
STEVEN'S DICTA

Stevens applied the following dicta to his experiments on estimation of sound decibels: beginning with a sound in the middle range to which the experimenter assigns a value, which the judge then uses a standard against which to compare subsequent stimuli; present variable stimuli that are both above and below the standard; vary the standard number in replications of the experiment with the same and different subjects; randomize the order of presentation of stimuli; let the judges present the stimuli to themselves, working at their own speeds; use 20–30 judges and work with the medians or geometric means of their estimates.

NOTES

[1] Crane. J.A. 1987. "Risk Assessment as Social Research", in PHILOSOPHY AND TECHNOLOGY III: TECHNOLOGY AND RESPONSIBILITY, edited by Paul Durbin. Boston/The Hague: Reidel, pp. 279–308.
[2] *Ibid.*
[3] Crane, "Risk Assessment . . ." p. 281. See note 1.
[4] Shrader-Frechette, K.S. 1985. SCIENCE POLICY, ETHICS, AND ECONOMIC METHODOLOGY. Dordrecht/Boston/Lancaster: Reidel p. 291.
[5] Lodge, M. 1977. MAGNITUDE SCALING. Beverly Hills, CA: Sage.
[6] E.g. measures of distributive justice. See W.M. Ives. 1982. "Modelling Distributive Justice Judgments" in MEASURING SOCIAL JUDGMENTS, edited by P.H. Rossi and S.L. NOCK. Beverly Hills, CA: Sage, pp. 205–234.
[7] in the JOURNAL OF SOCIAL SERVICE RESEARCH, 7(1), Fall 1983, pp. 81–92.
[8] *Ibid.*, pp. 84–85.
[9] *Ibid.*, p. 86.
[10] *Ibid.*, p. 88.
[11] *Ibid.*, p. 87.
[12] For a recent critique see Shrader-Frechette. 1985. SCIENCE POLICY, . . . note pp. 73–74.
[13] Mackenzie, B.D. 1977 BEHAVIOURISM AND THE LIMITS OF SCIENTIFIC METHOD Atlantic Highlands, N.J. Humanities Press.
[14] Mackenzie, BEHAVIOURISM . . chapter IV.
[15] *Ibid.*
[16] Crane, "Risk Assessment . . . ". See note 1.
[17] *New York Times*, November 9, 1985, letter to the editor by State Representative Bill Green from the 15th district.
[18] *Ibid.*
[19] *Ibid.*

[20] Crane, J.A. 1982. THE EVALUATION OF SOCIAL POLICIES. Boston: Kluwer-Nijhoff, p. 151.

[21] Crane, 1987, "Risk Assessment. . . . " (note 1) p. 298.

[22] Crane, J.A. 1976. "The Power of Social Intervention Experiments to Discriminate Differences Between Experimental and Control Groups". SOCIAL SERVICE REVIEW 50:2 pp. 224–242.

[23] *Ibid.*

[24] Probabilities of falsely finding a difference between experimental and control groups.

[25] Probabilities of failing to find such a difference.

[26] This term is discussed Shrader-Frechette, SCIENCE POLICY . . . (note 4) p. 68.

[27] A proposal for such a tribunal is presented in Shrader-Frechette, "SCIENCE POLICY . ." (Note 5), chapter 9.

[28] S.S. Stevens. 1966. "A Metric for the Social Consensus" SCIENCE 151:530–41.

[29] *Ibid.*

[30] Hamblin, R. 1975. "Magnitude Measurement and Theory", in MEASUREMENT IN THE SOCIAL SCIENCES, edited by H.M. Blalock, Jr. Chicago, Ill: Aldine pp. 61–120.

[31] *Ibid.*, p. 91.

[32] *Ibid.*, p. 115; Lodge (see note) p. 32.

[33] The experimental procedures became standardized, based on dicta developed by Stevens. See Appendix A.

[34] Lodge, MAGNITUDE ESTIMATION (see note) p. 24.

[35] *Ibid*; see also Hamblin, "Magnitude Measurement . . ." (note 30) pp. 65–69.

[36] Later developments are presented in Wegener, B. (ed) 1982. SOCIAL ATTITUDES AND PSYCHOPHYSICAL MEASUREMENT. Hillsdale, N.J.: Lawrence Erlbaum Associates.

[37] Hamblin, "Magnitude Measurement" (note 30) p. 80.

[38] *Ibid.*, p. 82.

[39] See note 35.

[40] Thompson, M.S. 1980. BENEFIT-COST ANALYSIS FOR PROGRAM EVALUATION. Beverly Hills CA: Sage, pp. 7–20. Hereinafter referred to as "BENEFIT-COST".

[41] Thompson, "BENEFIT-COST" pp. 175–181; Mishan, E.J. 1976. COST-BENEFIT ANALYSIS. New York: Praeger; Sugden, R. & A. Williams. 1978. THE PRINCIPLES OF PRACTICAL BENEFIT-COST ANALYSIS. London: Oxford University Press.

[42] In SCIENCE POLICY . . ., especially pp. 152–209 (see note 4).

[43] Shrader-Frechette, SCIENCE POLICY (note 4) pp. 194–200.

ALFRED NORDMANN

FUSION AND FISSION, GOVERNORS AND ELEVATORS

When James Watt introduced the governor as a feedback-mechanism that would regulate and control the operation of steam-engines, he established or consolidated a paradigm or exemplar for safety-design, for the liberal idea of political checks and balances, for the epistemological notion of self-correcting methodologies, a paradigm or archetype of intelligent systems, and the patron-saint of cybernetics.

The governor's privileged role as such a culturally central metaphor can be questioned by constructing and juxtaposing two technological models of safety, and by showing how the governor apparently instantiates only the weaker of the two. Like all other feedback-devices, the governor introduces a degree of safety into inherently unsafe systems as, for instance, today's standard commercial fission-reactor. The availability of the feedback-option distracts from the technical possibility of inherently and intuitively safe systems like the elevator or the envisioned fusion-reactor, both of which do not seem to be modeled upon a technological icon as powerful as the governor.

This claim will be established and explored by departing from a few somewhat puzzling features concerning the perception of technologies and implied attitudes towards safety. The attempt to account for these features leads to the development of a technological exemplar that can serve as the proposed alternative to the governor and other feedback-devices. This approach can be likened to the construction and juxtaposition of 'ideal types' as that term was used by Max Weber. Just like Weberian 'ideal types', the technological exemplars or models of safety represent one-sided generalizations (from the point of view of value-relevance) that guide hypothesis-formation. That is, they do not represent a complete or incomplete classificatory taxonomy designed to accommodate all technological systems. The specific merits of these ideal types and the specific ways in which they can guide hypothesis-formation will be discussed at the end of this somewhat speculative sketch.

A puzzling feature. Though tremendous amounts of energy are involved in and produced by fission and fusion reactors, though both

81

Edmund F. Byrne and Joseph C. Pitt (eds), Technological Transformation: Contextual and Conceptual Implications, 81–92.

technologies require highly complex mechanical apparatus prone to material weaknesses, though both operate on the subatomic level in the organization of matter, and though both depend on a tenuous process of translation from theoretical physics to technological application, fusion reactors are presented, perceived, and developed as a safe alternative to fission technology. In order to understand just what structural dissimilarities correspond to the labels 'safe' and 'unsafe' in respect to these technologies, it may be helpful to develop formal characterizations for the governor, on the one hand, the elevator on the other. The decisive difference between these two systems can be seen to extend to fusion and fission and, by implication, to a number of other technologies.

Two puzzled questions. Why did the governor inaugurate the era of automatic control, and with it, the field, if not the science of control engineering? And why did such a seemingly simple and uninvolving idea as transporting people vertically on platforms or in cabins experience a sudden breakthrough only after 1853?

As everyone knows, a steam-engine pressurizes steam in a boiler, and that steam sets into motion some other sort of machinery. Governors were used to regulate the speed of motion produced by the emission of the pressurized steam, but Watt also used it to regulate boiler-pressure, i.e. for preventing a build-up of pressure in the boiler that could lead to an explosion. In this function, the governor serves most straightforwardly as a safety-device. More particularly, Watt's centrifugal governor would be driven by the steam – the greater the pressure in the boiler, the faster the governor would rotate around its own axis to which are attached flyballs which extend further outwards as the speed of rotation increases. Once centrifugal force extends the flyballs outwards to a predefined height, an attached mechanism will open a valve, steam will be released, the pressure in the boiler will decrease, the governors rotational motion slow down, the flyballs sink again and the valve close – until sufficient pressure has built again and the whole process can start all over.

The governor thus intervenes between two of the overall system's three states. The overall system can be in a state of rest or an *arrested* state (no heat, no pressure, no work performed), in an *operative* state (steam is produced and the machinery ready for operation), and in a *rampant* state (out of control, explosion due to too much pressure). The operative state is defined as the normal state for the system. Indeed, the system is ingeniously engineered for efficiency, ensuring the greatest possible stability of the operative state. Since, left to itself, the system would tend from the operative towards the rampant state, the governor is to serve as a buffer and mediate between the rampant and the operative states such that the rampant state will never be actualized. While the system maintains and stabilizes itself in the operative state, counteracting the tendency towards the rampant state, the system cannot bring itself into or out of the system's first and final state, the arrested state. It performs no function, and only a – one could say – managerial decision can move the system into or out of that state, i.e. the decision to shut down the system or re-initiate operation. Some feedback-controlled systems (life fission reactors) include an automatic shut-down capacity (by flooding the reactor, for instance). Here, a managerial decision is included within the system's competence, a kind of self-destruct feature as far as the system's self-regulatory capacity is concerned: the system returns itself into managerial hands.

This system is called safe insofar as its tendency to move from the operative to the rampant state, i.e. from safe performance to catastrophe, can be successfully checked by a technology which monitors and regulates the system's performance. The system is safe when its monitoring and regulating technology is perfect, when it completely succeeds to stabilize the normal or operative state, to insulate it against all "external disturbances" (Mayr, 1970, p. 8). And ever since the discovery of oscillations induced by the steam-engine's governor, control engineering moves towards that goal of perfection as the regulatory technology gets more and more sophisticated. But the goal of perfection cannot be fully realized as long as time lapses until changes in the system are registered and regulatory moves implemented, as long as the regulatory technology possesses material qualities of its own which affect its performance, possibly the system's overall performance, or the performance of other regulating devices. According to the governor's exemplary status as the paradigm of safety-devices, we have grown accustomed to use the predicate 'safe' as a relational predicate: nothing is absolutely safe. Safety has become a matter of judgment or degree, in either case depending on the probability and severity of an accidental realization of the system's rampant state.

The Otis-elevator, on the other hand, provides an example of a system which can be said to be absolutely safe, insofar as it is a two-state system encompassing only the arrested and the operative state, a system for which no rampant state can be defined and which, accordingly, possesses no tendency to move towards such a rampant state or catastrophe. This is not to say that one cannot get hurt using an elevator or that there is no preconception as to what the rampant state would be for this kind of technology. Indeed, and quite to the contrary, there is an intuitively all too compelling vision of what that rampant state would be for the elevator: a freely falling cabin. Also, people did and do get hurt and killed in elevator-accidents, especially maintenance personnel. So what sense does it make to say that elevators are absolutely safe?

A clue. In 1853, Elisha Graves Otis gave a dramatic public display of his newly designed elevator at New York's Crystal Palace. Standing on the elevator's platform, Otis had himself elevated to considerable height in full view of an onlooking crowd, an assistant then cut the cable on which the platform was suspended – the anticipated fall did not take place. And this is how it worked: the cable held the cabin at the joint of a spring so that the force of the cabin's weight against the cable would contract that spring and allow for free up- and downward motion of the cabin in the shaft. Thus, if the cable should break, the force on the spring will instantly cease and the spring consequently snap outward and lock the cabin into a toothed rail in the shaft. (See Fig. 1, p.90).

This system has no auxiliary safety-device, there is no feedback-loop that would monitor and regulate the elevator's safe performance. Also, it can be in only one of two states: either it sits locked into the shaft without support from the cable, clearly in an arrested state, or it is suspended on the cable, contracting the spring, thus in its operative state regardless of whether it actually is or is not in motion. Both states are safe, and in this kind of system, the arrested state is defined as the normal state, while the operative state is tenuous or fragile: it is conditioned upon the strength of the cable which should be strong for the purposes of efficient operation but need not be strong for safety-purposes. Here, the stability of the operative state is solely an economic virtue, the notions of stability, safety, and efficiency are not conflated with one another. Also, the energy needed to place the system into its operative state is not pooled with the energy that will produce and is somehow proportional to the system's performance or output. On the contrary, the system is tenuous or fragile insofar as it is moved from its normal, arrested state into the operative state only if a constant force is established and maintained throughout operation. Once that force drops out, the system reverts into its normal state of arrest. As in a mousetrap, the energy needed to place the system into its operative state becomes transformed into stored up, suspended, or latent energy, waiting to return the system into its arrested state. And the arrested state obtains not when the system does not work but when it cannot work. All this quite evidently sets the safe system 'elevator' apart from the unsafe systems 'steam-engine' or 'fission-reactor'. While they also need energy to induce the system's operative state, energy from the same source and through the same channels is needed to ensure actual operation. Also, these systems require further energy to either quench the fire, flood the reactor or move the control rod.

I propose that the difference between the two systems can be used to define 'safety'. It is the difference between a two-state system, consisting of a normal arrested state and a tenuous operative state, and a three-state system, consisting of an arrested state and a normal operative state tending towards a rampant state. It is the difference between a system that reverts to its arrested state as the specific force needed to maintain the operative state drops out, and a system requiring control technology to intervene between the operative and rampant states and to introduce degrees of safety into the inherently unsafe system. It is the difference between a system as fragile as a mouse-trap, and a system in which safety is predicated upon an economic and only then technological virtue, stability. Freeman Dyson (1979, p. 98) calls it the difference between "'engineered safety,' which means that a catastrophic accident is theoretically possible but is prevented by the way the control system is designed" and 'inherent safety,' [. . .] guaranteed by the laws of nature and not merely by the details of its engineering."[1] This is not to claim that the fall of an elevator-cabin is a physical impossibility or that other kinds of elevator accidents (e.g. people falling between the cabin and the wall of the shaft) will not occur. For instance, one can easily imagine a snapping cable and material fatigue of the spring or the toothed rail resulting, after all, in a falling cabin. The absolute notion of safe and unsafe systems cannot replace the relative notion of safety as a question of likelihood and severity of accidents, but it precedes it: the likelihood and preventability of accidents in inherently safe two-state systems should be approached and assessed differently than the likelihood of accidents in three-state systems engineered for safety. So-called fault-tree analysis, for instance, may be appropriate only for two-state systems since the structure of a branching tree does not capture the flow of events in a system characterized by feedback-loops. While the structural weakness of certain component parts may be the only source of failure in two-state systems, the introduction of feedback-devices not only heightens the system's opacity but leads to a proliferation of potential sources of failure which now include operator error, the so-called 'miraculous' event which overcomes all functioning regulatory controls, failures in particular feedback devices, and failures generated by the interaction of various feedback loops. Especially these latter two potential sources of failure increase with the number of implemented controls, the so-called redundancy of back-up systems thus transformed into a hazard of its own. The greater difficulty of assessing the risks posed by these systems is further illustrated by a problem commonly faced by computer programmers. While they can motivate the introduction of each loop, routine, or subroutine into their programs, they create an overall system so opaque that its logical structure becomes inscrutable even to themselves and can be checked only through countless randomized test-runs which will hopefully reveal the program's flaws. All this, in turn, helps justify and explain the fear of accidental nuclear war. Even before the public learned, for instance, that a malfunctioning centrifugal feedback sensor was largely accountable for near-catastrophe at Three Mile Island, the 1963 movie "Dr Strangelove or How I Learned to Stop Worrying and Love the Bomb" showed how a backup procedure implemented as a safeguard may provide an opportunity to launch an unwarranted nuclear attack.

This conceptualization of 'safety' accounts for the use of that term when fusion-reactors are presented as a safe alternative to fission-reactors. While the H-bomb shows that the physical process of fusion is not intrinsically 'friendlier' or safer than fission, the physical constraints upon controlled fusion and fission for civilian purposes have led to fundamentally different design-proposals. Today's commercial fission-reactors require elaborate feedback-loops to prevent a melt-down and a release of high levels of radioactivity. Once induced, a fission-process can be halted only through active interference, including the dropping of the safety rods. Envisioned fusion-reactors (like the Tokomak reactor), on the other hand, possess only a most tenuous and fragile operative state as the fusion-process can only take place at extraordinarily high temperatures under precarious conditions of containment. Just as nothing can go wrong with an elevator but the breaking of a cable, so with these fusion-reactors: anything that might go wrong will first of all affect the temperature necessary to maintain operation; as soon as that temperature drops, the reactor reverts into its safe arrested state. Again, no rampant state can be defined for fusion reactors.

The proposal and defense of fusion-reactors thus rests on a notion of 'safety' that differs significantly from the use of that term in the context of control-engineering as far as that discipline departs from the governor as a model for safety-devices. Having identified the alternative uses of the term 'safety' and two ideal types of technical systems, we should ask how all this contributes to our understanding, increases our options, or advances hypothesis formation. Two cautionary remarks should be made and six different ways of extending this sketch can be suggested here.

As a first cautionary remark it is important to note that one cannot claim the wholesale superiority of two-state systems over feedback systems. There are tremendous differences in the design of two-state systems as illustrated by the juxtaposition of dead man's throttle and car-accelerator. Also, a two-state system that is badly designed or built may pose far more hazards even than a fairly unsophisticated feedback-supported technology. The second proviso pertains to the taxonomic qualities of the ideal types. Most larger technological systems encompass components of both types and even individual components may not neatly exemplify either type. Mayr, for instance, includes Papin's weight-loaded valve among feedback devices though admitting that it does not meet his third criterion of feedback-systems (namely that either the sensing element or the comparator is physically separate – Mayr, 1970, p. 8). Though possibly less plausibly, however, this kind of valve can also be interpreted as constituting an inherently safe two-state system.

Beginning now with the most naively stated of the six ways of extending this sketch, the ideal types of technical systems may serve as a design heuristic. For each feedback-based three-state system, can one develop a safe two-state system that will perform roughly equivalent tasks with roughly the same efficiency?

System theorists and physicists would be called upon to explore the differences between the two kinds of systems. Which of the various distinctive traits discussed here represent necessary or sufficient characteristics of safe or unsafe two- or three-state systems, how are these traits interrelated, and do these systems invite (as is currently fashionable) thermodynamic characterizations as more or less integrated systems, sensitive to greater or smaller energy-fluctuations?

Thirdly, historical and sociological investigators of attitudes towards technology may extend the implied explications of 'safety' to form explanatory hypotheses about patterns of acceptance and rejection of new technological products. For instance, why do we feel more comfortable towards biodegradable compounds than towards compounds that are to be detoxified or reintegrated through some process of 'recycling'? Can one say, perhaps, that biodegradable compounds (which are chemically stable only in the environments in which alone they are supposed to persist) belong into an inherently safe two-state ecological system while detoxification and recycling plants function like feedback-loops in a safety-engineered three-state ecological system?

There may be certain implications for the political, economical, and technological problem of international technology transfer. The juxtaposition of two-state and feed-back based systems lends conditional support to a number of familiar criticisms that have been leveled against the export of certain high technologies. These technologies are to create new dependencies, on the one hand in terms of spare parts, repair and implementation skills, on the other hand as a dependency on an entire body of scientific knowledge (e.g. on the science of control engineering). With these technologies are to come particular uninvited constraints on the rational organization of work, i.e. a particular mode of management involving a hierarchy of operators depending on the degree of required oversight over the whole technology. Furthermore, these technologies may inadvertently realize values that are alien to the culture into which they are imported (equating, for instance, safety with stability or alleviation of physical labor with reduction of the workforce). Finally, high technology is supposedly too opaque, not robust enough, too expensive, and sometimes too dangerous for users in less 'developed' countries. All these criticisms may pertain primarily to feedback-based technologies and differently or not at all to two-state systems.

Philosophers of science and technology may find in the two ideal types an instructive example of how extraneous values pervade technological artifacts without having been placed there deliberately. In the case at hand, this is most obvious for the supposed mutual dependence of safety and stability which is often taken to be something of a technological truism. On closer inspection, however, stability may turn out to be an economic virtue without any necessary relation to 'safety'. And yet, the engineer perfecting the stability of a system's operative state may well do so only in the name of 'safety'. While unwittingly perpetuating the 'ideology' of safety–stability, that engineer can be seen to 'really' advance economic interests.

Finally and in a most philosophical vain, we may begin to speculate how a non-feedback-oriented, absolute notion of safe systems affects our understanding of intelligent systems, or how it may challenge the governor's status as a central metaphor in the political, technological, and intellectual culture of our time. For instance, it has been shown that the genesis of political systems of checks and balances, e.g. the balance of the executive, legislative, and judicial powers, was closely related to the development of the governor, of control engineering and its promise of safety or security. Alexander Hamilton, for instance, recommended (as quoted by Bennett, 1979): "Make the system complete in its structure, give a perfect proportion and balance to its parts, and the power you give it will never affect your security." Could there be an alternative notion also of social or political security, a two-state system in which, for instance, no power is invested into any branch of government but where government is performed by a legislature all members of which are subject (individually or as a whole) to immediate recall at any time?

ACKNOWLEDGEMENTS

I would like to thank two anonymous referees, Richard Burian, and Carl Mitcham for their stimulating comments and hints.

NOTES

[1] See Diagram 1 and Diagram 2, p.91.

Fig. 1. A demonstration in 1861 of Elisha Graves Otis's "Ratchet Safety Platform." The anticipated fall does not take place. As the general anxious expectation of the rampant state is frustrated, the operator triumphantly exclaims: "All safe!" (From Simmen and Drepper, 1984, p. 13.).

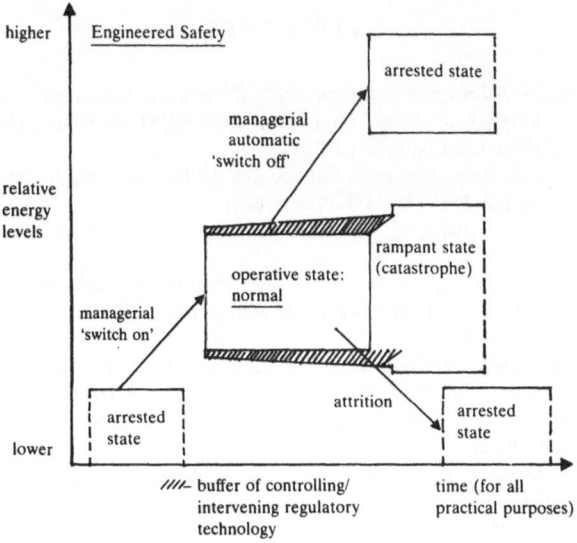

Diagram 1.   Engineered safety. Only the duration of the 'operative state' is bounded at both ends by various well-defined events.

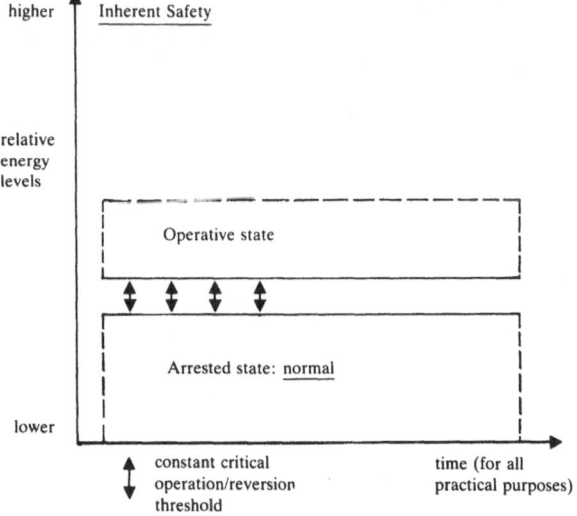

Diagram 2.   Inherent safety. The system is in its arrested state not when it does not but when it cannot operate. And the system's actual performance or the production of some output proportional to some input is here only a particular event within the system's operative state.

REFERENCES

Automatic Control. A Scientific American Book. (New York: Simon and Schuster, 1955).

Stuart Bennett. *A History of Control Engineering 1800–1930* (London and New York: The Institution of Electrical Engineers, 1979).

Joan Lisa Bromberg. *Fusion: Science, Politics, and the Invention of a New Energy Source.* (Cambridge and London: The MIT Press, 1982).

David Dickson. *The Politics of Alternative Technology.* (New York: Universe Books, 1971).

Freeman Dyson. *Disturbing the Universe.* (New York: Harper and Row, 1979).

David Rittenhouse Inglis. *Nuclear Energy. Its Physics and its Social Challenge.* (Reading: Addison-Wesley, 1973).

Friedhelm Korte et al. *Oekologische Chemie. Grundlagen und Konzepte fuer die oekologische Beurteilung von Chemikalien.* (Stuttgart, New York: Thieme, 1980).

Thomas Kuhn. *The Structure of Scientific Revolutions.* Second Edition. (Chicago: University of Chicago Press, 1970).

Jacques Ligou. *Elements of Nuclear Engineering.* (Chur, London: Harwood, 1986).

William Lowrance. *Of Acceptable Risk. Science and the Determination of Safety.* (Los Altos: Kaufman, 1978).

Otto Mayr. *Authority, Liberty, and Automatic Machinery in Early Modern Europe.* (Baltimore and London: Johns Hopkins University Press, 1986).

Otto Mayr. *The Origins of Feedback Control.* (Cambridge and London: MIT Press, 1970).

Jeannot Simmen and Uwe Drepper. *Der Fahrstuhl. Die Geschichte der vertikalen Eroberung.* (Muenchen: Prestel, 1984).

Max Weber. "'Objectivity' in Social Science and Social Policy." In: M.W. *The Methodology of the Social Sciences* (New York: Free Press, 1949), pp. 49–112.

JAMES KLAGGE

THE GOOD OLD DAYS:
AGE-SPECIFIC PERCEPTIONS OF PROGRESS

> "Progress might have been alright once
> but it has gone on far too long."
> *Ogden Nash*

I

Codger: Remember the good old days? You could get an ice cream cone for 5 cents. You knew all your neighbors and they'd help you out when you needed help. People worked hard, but they appreciated what they had. You didn't have so many complicated choices to make. Life was simpler then. Those were the good old days.

Whipper-snapper: You must be kidding! Sure you could get an ice cream cone for 5 cents, but you had to hoe corn from sun-up to sun-down just to get 25 cents. That didn't leave much time for anything else. Anyway, there was no entertainment to enjoy. You had to wash up in cold water, go outside to use the toilet, and the only way to stay warm in winter was to go outside and chop wood. You had no choice of what to do in life. You grew up believing what your parents believed, you did what your father had done for a living, you married the girl next door, and you had more children than either of you wanted.

Codger: I know, but I was happier then. Life was simpler.

Whipper-snapper: You're just getting senile. Your memory is operating selectively. You're blocking out the limitations and hardships, and focussing on the few good things. You think you were happier then because you are only remembering the happier times. And you are comparing it with now, when you are aware of unhappy times.

Codger: Perhaps. But I think I'm being fair. I remember all those bad things you mentioned, but they just didn't matter much.

93

Edmund F. Byrne and Joseph C. Pitt (eds.), Technological Transformation:
Contextual and Conceptual Implications, 93–104.
© 1989 Kluwer Academic Publishers.

Whipper-snapper: Look, you're older now, and older people are less happy than younger people. You were in the prime of your life then. You lived with zest. Now you are old. You feel useless and unwanted. Most of your old friends are dead. Your body is deteriorating, so you experience limitation and pain. If you had been old then, you would have been unhappier (as you are now), and if you were young now, you'd be happier (as you think you were then).

Codger: Again, you may be right, but I'm trying to be fair. I knew old people then. They lived with their children, they did what they could around the house, and they died with dignity. Old people now live alone, or in homes with other old people we don't know. We die in hospitals, alone, usually long after we are of any use to ourselves or to others. We live longer, but we are no better off for it. Our bodies usually end up outliving our minds. I know young people now. But they don't seem to get any satisfaction out of life. They have a lot more than young people used to, but they don't seem to enjoy what they have very much, and they always seem to want more than they have. They worry about their job, their physical fitness, and their cars. Nothing satisfies young people any more. In the good old days young people were satisfied.

Whipper-snapper: If you could take a time-machine back to those so-called good old days, knowing how things are now, you would stop thinking they were so good. People were probably not as satisfied then as you think, but to the extent that they were satisfied then, it was because they didn't know any better. If people have limited ambitions and options, it is easier to satisfy them. But no one would want to go back and limit his options.

Codger: You're right, but I'm not sure what that proves. I wouldn't be happy if I returned in a time-machine, knowing what I know now, but that doesn't show people weren't happier then. I think people were happier then than they are now – I certainly was. And I am trying to be conscious of all the warnings you properly made. Why are you so sure I'm wrong? You didn't live then.

Whipper-snapper: I guess I have always pictured life in the old days from the perspective of my own ambitions and options. From that perspective it looks pretty dreary. But, of course, fortunately for them, people who lived then didn't have my perspective. So you are not advocating that we throw away our technology and social progress and return to the good old days?

Codger: Heavens, no! We would never be happy with that.

Whipper-snapper: Then what are you getting at?

Codger: I'm just afraid that someday people will look back on 1987 as the good old days.

Whipper-snapper: Perhaps some will. But only the old and senile. And no one will pay attention to them.

II

Let us reflect on what Codger has said. What is the relationship between technological growth and human well-being? The following graphs indicate some of the main possibilities:

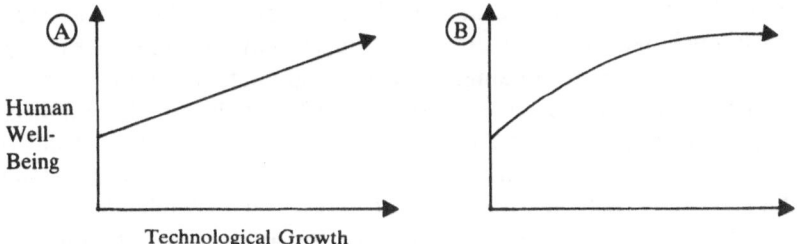

Graphs A, B   Graphs A and B represent human well-being as a monotonically increasing function of technological growth. In the case of B, human well-being has an upper bound (with respect to technological growth).

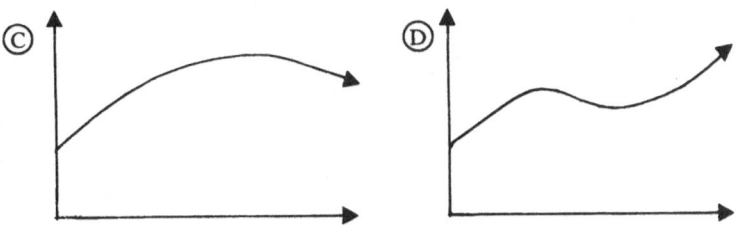

Graphs C, D.   Graphs C and D represent the occurrence of inverse correlations over certain intervals of technological growth. C represents the inverse correlation as permanent after a certain level of growth. D represent the inverse correlation as temporary, though possibly recurrent.

Codger may lead us to think that human well-being is sometimes inversely correlated with technological growth, as represented by graphs C and D. Codger's position involves or requires a number of qualifications:

1. Technological growth is clearly not the only factor that influences human well-being. Other factors, such as social, political, and economic arrangements, religious and moral climate, and evolutionary factors, may be significant in influencing human well-being.

The extent to which this is a serious qualification depends on the way in which we define "technology". Suppose we define technology as: *any artifact that increases our ability to achieve our ends.* According to this definition, social, political, and economic arrangements could be instances of technology.[1] Of course we don't want to make the concept of technology so broad that we are virtually stipulating that technology is the only influence on human well-being. As I have defined technology, some relevant factors that are not technological are the availability of natural resources, and the prevalence of and our susceptibility to diseases. Yet these in turn seem to be largely influenced by technology.

Although technological growth is clearly not the only factor that influences human well-being, it seems legitimate to treat it as the fundamental influence, in terms of its direct as well as indirect effects.

2. Codger has not attempted to distinguish the effects of various individual technologies on human well-being – a task that would be rather difficult. It is quite likely that some individual technologies are not inversely correlated with human well-being. Codger simply lumps all of technology together in its effects.

3. Codger's claim is limited to the circumstances of his own lifetime. He is not making a claim about people who are in vastly different circumstances of affluence or technological development. There are bound to be circumstances of technological growth – perhaps even most of them – that do not confirm Codger's experience.

4. Codger is not appealing to any of the standard complaints about technological growth – that it leads to deterioration of the environment, depletion of natural resources, catastrophic accidents, and health harms. Let us call these possible effects of technological growth "negative direct effects". Complaints about these may have some substance to them, but they are apparently not what Codger is complaining about.

5. Finally, Codger reports only his own experience. Other people who

lived through the same time and circumstances might feel differently. But Codger's experience does not seem to be uncommon, at least among old senile people.

Despite all of these qualifications, Codger's position is an interesting one. It is interesting because it is paradoxical. If we understand "technology" as any artifact that increases our ability to achieve our ends, and if we ignore negative direct effects, it seems to be a sheer truism that technological growth improves human well-being.

III

In what ways, besides negative direct effects, might technological growth adversely affect human well-being? There seem to be three general ways. Let's call these Psychic Effects:

1. Technology can create ends for us by making things possible that were not previously possible.
2. Technology can decrease the satisfaction we get in achieving our ends.
3. Technology can depreciate certain ends by making other ends more attainable and, hence, more attractive.

All three of these ways trade on the fact that means can themselves influence ends. Let us examine these in turn.

1. Since pursuing and achieving ends is a main source of human well-being, one might think that creating ends could only improve human well-being. But acquiring ends is only the first step in pursuing and achieving them. One can acquire ends without having or acquiring the ability to pursue or achieve them. It is, of course, irrational to acquire ends that one cannot reasonably hope to pursue or achieve. But that is a form of irrationality to which humans are commonly susceptible.[2]

Commercial advertising trades on the ability to create ends for people by making the ends *seem* achievable, or by deflecting the question of their achievability. But for many people the end is acquired while the means are still lacking. Of course, the well-being of some – those who can pursue and reasonably hope to achieve the ends created – may be increased, but to the extent that the means for achieving ends are unequally distributed, frustration will be more common than satisfaction. Since producers and advertisers are benefited by creating ends for people, and they are not harmed by frustration, there is an incentive to use technological growth in ways that end up decreasing human well-being overall.

This effect is not limited to the insidious effects of commercial advertising. Any communication of information about technological growth can have this effect because humans are susceptible to wishful thinking, even when no one tries to encourage this in them. This is one of the reasons we cringe at the thought of TV antennas on grass huts or tin shacks.

It is important to note that this negative effect on human well-being would not occur if humans were fully rational. If no one formed ends that were insufficiently attainable, then technological growth would not induce frustration. Technological growth, then, is not harmful *per se* in this respect. Rather, it is awareness of technological growth by less than perfectly rational people that is harmful. This makes it easy to think that the problem lies with people, not with technology – it is their own fault if they cannot handle it. And it is also easy for a man to think that he can handle it – it is those who rattle on about the "good old days" that cannot. But that is what is so insidious about this effect. It is difficult to detect it operating on oneself.

2. Technology can decrease the satisfaction we get from achieving our ends, in either of two ways. First, it can do so by making it easier to achieve the ends. It is commonly said, by old senile people, that one values more highly what one has worked hard for. If one's well-being is influenced by the value one places on the achievement of one's ends, and that value is decreased, then so is well-being.

Perhaps the decrease in value from a given achievement can be offset by achieving more ends. Indeed, since technology makes it easier to achieve ends, it should make it possible and easier to achieve more ends. Yet, one may have to contend with a diminishing marginal utility of achievement. In that case, it may be rather difficult to make up for loss of quality by increase in quantity.

Another respect in which technological growth can decrease the satisfaction of achieving ends is through the proliferation of possible ends. Let us assume all the possible ends are reasonably achievable by the individual in question. The problem lies in the fact that we cannot predict in advance the likelihood of, or satisfaction from, achieving the various ends. Thus, in choosing which ends to pursue, we are choosing under a large degree of uncertainty. To the extent that our limited energies force us to choose among competing ends, we will never know for sure how well things might have turned out if we had made different choices. This should be called the Grass-is-greener-on-the-other-side phenomenon. We rarely think how lucky we are that things turned out as well as they did. We tend to anguish over how much better they might have turned out than they did. Thus, we may get less satisfaction from the pursuit and achievement of our ends than we would have if those ends had been the only ones we had or could have had.

Decreasing satisfaction from easily achieved ends is not a manifestation of irrationality. But it is something that we regularly ignore when we make things easier for ourselves, even when we can identify its operation in our past.

Decreasing satisfaction from uncertainty over counterfactual outcomes can be a manifestation of irrationality if one can or will no longer change which end one is pursuing. On the other hand, it can be rational to rethink one's ends in light of information gathered from pursuing one of them. But since uncertainty about the satisfaction to be attained from the pursuit of other ends is so great, one may still be in no position to make a more rational comparative choice. Unless one's choice has turned out rather badly, little can usually be gained, and much suffered, from second thoughts. And it is technological growth that allows for the second thoughts.

3. By creating certain ends, and making certain ends easily achieved, technological growth tends to seduce people away from kinds of ends that are not easily achieved but are deeply satisfying to achieve. Technological growth tends to sap the importance, in people's minds, of cultivating interpersonal relationships and personal skills, and substitutes instead ends that bring immediate and easy satisfaction.[3]

This tendency is a manifestation of irrationality. We tend to prefer immediate satisfaction to later satisfaction even when we know the later satisfaction will be greater (or we tend to overestimate the uncertainty involved in the later satisfaction). This is known as pure time preference. Purely rational people would not be near-sighted in this way, as actual people tend to be. At any given moment we tend to care less about our overall well-being through time than about our short-term well-being. And technological growth caters to this weakness in us.

There is another more diabolical sense in which technology can influence the ends we pursue. Technology, especially advanced technology, increases the ease with which we can manipulate and dominate other people – through, for example, weapons, electronic surveillance, and mass communication. While it is true that technology also increases the ease with which we can help others, the fact is that humans are very susceptible to opportunities for advancing their own interests and controlling others. Technology not only does nothing to check this impulse, it constitutes a temptation for it. Again, technology, while not bad in itself, appeals to something bad in us.[4]

IV

But is Codger right? Though I haven't bothered to survey the ways in which technological growth improves human well-being, they are enormous. Could they really ever be outweighed by the negative effects, both direct and psychic? In early stages of technological growth they clearly cannot. Negative effects (especially psychic ones) begin to mount only when technology has grown considerably. How considerably is not clear. But the presence of Codgers suggests we might have already reached such a point in the United States.[5]

If there is a point at which negative effects begin to outweigh positive effects, is this a temporary state? And if it is temporary, is it also recurrent? People who "believe" in technology will believe that any dips in the graph are only local, and that over sufficiently large intervals the graph rises. In other words, technological growth can always overcome its own negative effects.

Old people like Codger don't believe in technology in this sense. It could be that old people are more sensitive to the subtle ways in which technological growth caters to human weaknesses. Since technological growth always presents itself as perfectly rational, it easily obscures from us its negative psychic effects. Who would think that something so rational caters to something irrational in us? Only someone with years of experience, who can discern its subtle effects on himself and on us. Only an old and, apparently, senile person.

Perhaps Aristotle was right when he recommended:

> Therefore we ought to attend to the undemonstrated sayings and opinions of experienced and older people or of people of practical wisdom not less than to demonstrations; for because experience has given them an eye they see aright.
>
> (*Nicomachean Ethics* VI, 11)

APPENDIX

Early explorations of the value and danger of technology can be found in the Greek myth of Prometheus and the associated myth of Pandora. The earliest and most interesting sources of the myths are Hesiod (Eighth Century B.C.) and Aeschylus (Fifth Century B.C.).[6]

Though there are differences between versions of these myths, the basic structure of the story is this: Prometheus stole fire from the gods and gave it to mankind for its benefit. Zeus sent punishment in return.

A. The Prior Settings

In Hesiod's *Works and Days* mankind earlier had had fire and other technologies and lived in an idyllic state, a day's work producing a year's requirements. Prometheus, in an attempt to outwit Zeus, tricked (or tried to trick) Zeus into accepting oxen bones and fat as the proper sacrifice, reserving the edible parts for man. Zeus was angered by this and took fire away from man as punishment. Prometheus stole it again for man.

According to Aeschylus, on the other hand, mankind had been in a very beastly and precarious state, and Zeus was about to blot it out. Prometheus stole fire, which mankind had never previously had, to benefit mankind and save it from extinction.

B. The Punishments

In Aeschylus' play, Zeus punished Prometheus for stealing fire by having him staked to a mountainside by Hephaistos, the lame god of technology. In *Works and Days*, Hesiod has Zeus send the punishment to mankind in the form of the woman, Pandora, who was created from clay by Hephaistos. Pandora released all evils into the world by illicitly opening a jar containing them. In the *Theogony*, Hesiod includes both these punishments, of Prometheus and of mankind, though the punishment of mankind is simply woman herself, symbolized by Pandora, and not also the released contents of a jar.

C. Some Meanings

Though these myths reflect many concerns, they are clearly concerned with technology. Fire exemplifies technology and symbolizes a wide range of technology and technical knowledge. This wide range is spelled out by Prometheus in Aeschylus' play. What views about technology do we find in this conglomeration of myths?

Aeschylus' Prometheus sees technology as an unmixed blessing for mankind. According to the play, mankind receives no punishment for receiving fire from Prometheus. Indeed, Zeus does not blot out mankind after all, but directs his anger against Prometheus instead.[7] Prometheus himself suffers from man's possession of technology, but he continues to "believe" in technology. He sees his own punishment as an excessively capricious exercise of Zeus' newly-gained power.

But it is not clear that Aeschylus' play confirms Prometheus' view of technology. If we see Prometheus as symbolizing mankind in the play,[8] or as symbolizing the divine element in man that attempts to overreach his proper condition, the play can be seen as reacting to the common view, espoused by Prometheus, that technology is the salvation of mankind. The play may be expressing the truth, embodied in the punishment of Prometheus, that we easily miss seeing the dangers of technology, even though they are right before our eyes. We miss seeing how it is that we suffer from technology.[9]

In *Works and Days*, Hesiod does not describe the condition of pre-technological man, as Aeschylus had.[10] He describes a life of ease lost by Zeus' hiding fire. Man's loss of fire seems gratuitous since it is caused by Prometheus' trickery, unless, again, we identify Prometheus with (a divine aspect of) mankind. Fire is then stolen by Prometheus and returned to mankind, but only at a price. In Hesiod's version the price is paid by mankind, through the introduction of evil into the world.

As with Aeschylus' play, Hesiod's myth makes the connection between technology and human harm appear gratuitous at one level. Indeed, man had lived idyllically with technology at an earlier time. But it is possible to see the myth as embodying the thought that technology has its price, even though it is not easy to see the connection. The fact that there is a span of time between the initial acquisition of technology and the price can be seen in two ways. The time span can be understood literally – technology eventually has its price – or it can be understood metaphorically – technology seems on the surface to be an unmixed blessing, but a deeper examination shows it to be fraught with evils. Technology gives man the impression that he can achieve anything, even fool the gods. Technology is pure means to any ends we choose. But the myth suggests this faith in technology becomes a disguised danger. Pandora, made by the god of technology, comes as a beautiful woman, bearing evils in her soul and in a jar. In the *Theogony*, woman (like technology) is an evil we cannot happily live with, yet we cannot happily live without it either.

The Greeks were always sensitive to the dangers of *hubris* – over-reaching oneself, thinking of oneself as more than merely human. They saw technology as making man very susceptible to *hubris*.[11] They were inclined to attribute the cost to the jealous wrath of the gods. From a modern perspective, it is possible to see the danger of technology in the disturbance of the human situation – whether it be disturbance of the human *psyche* (as I have emphasized in discussing the psychic effects of technology), or of the human ecology (as others are inclined to empha-size in discussing what I have called negative direct effects of technol-ogy). It is, I think, hubristic to ignore the fact that we are partly irrational (i.e., merely human). Technology is dangerous precisely because it conceals its negative effects on well-being through its seduc-tive appeal to the irrational aspect of the *psyche*.

The Ancient Greeks, as I read these myths, betray a premonition of these hidden dangers of technology.[12]

James C. Klagge
Virginia Polytechnic Institute and State University

NOTES

[1] One might wish to limit technology to physical artifacts, in which case it may not include such institutional arrangements.

[2] Jon Elster has emphasized the value of a certain gap between ends and prospects for achieving those ends for getting people to stretch their abilities. He calls this an optimum level of frustration. See Elster, "Sour Grapes – Utilitarianism and the Genesis of Wants," in A. Sen and B. Williams (eds.), *Utilitarianism and Beyond*, Cambridge, 1982, p. 233. Of course, some frustration is not productive, as when Latin Americans take their ideal of female beauty from blonde, fair-skinned images.

[3] Marx would have to hold that this is due to capitalistic alienation. Under communism Marx holds that advanced technology will leave significant free time, which he expects will be devoted to the all-round development of the individual. See, for example, the passages from Marx's *Grundrisse* excerpted in D. McLellan (ed.), *Karl Marx: Selected Writings*, Oxford, 1977, pp. 368–387.

[4] Others would explain Codger's attitude toward the "good old days" as resulting from

the increasing distance between his expected levels of achievement and his actual levels of achievement. (This "rising expectations" explanation is given by Nicholas Rescher, in "Technological Progress and Human Happiness," in *Unpopular Essays on Technological Progress*, Pittsburgh, 1980, pp. 12–14. Cf. also Elster, pp. 231–236.) I think this explanation is fine as far as it goes (indeed, I include a form of it in my discussion of Psychic Effect 1 in Section III, *supra*), but I hope my explanation has gone somewhat deeper. Rescher's explanation is predicated on the assumption that rates of actual achievement decelerate over time. Only then can projection of past rates of achievement be certain to overestimate actual future rates of achievement. First, I am not sure we can assume deceleration. Second, I think well-being can suffer even in the absence of deceleration, as indicated by Psychic Effects 2 and 3.

[5] *Upstart*: But there have always been Codgers around. Despite his disclaimers, it seems likely that the explanation of his view of the past has more to do with human nature than with the nature of technology. People, especially old people, tend to adore the past, regardless of the levels of technological development.

[6] See Hesiod, *Works and Days*, 42–104; *Theogony*, 508–616; and Aeschylus, *Prometheus Bound*, 1–560.

[7] It is not made clear why Zeus does not blot out mankind after all, despite its possession of fire. Could it be that, in the Aeschylean story, fire has made man invulnerable to divine whim? This seems unlikely. Yet technology may well make us think we are invulnerable.

[8] The same identification is useful for explaining Zeus' punishment of mankind for Prometheus' trickery.

[9] It is important to make clear that in reading certain views into these myths, I am not claiming that these views were consciously espoused by Hesiod or Aeschylus. My claim is that the myths embody certain primordial insights about technology – that it increases our susceptibility to *hubris*, and that *hubris* has its cost. These insights get articulated by Hesiod and Aeschylus in a religious sense, but I think the insights can be understood quite naturalistically.

In "Prometheus Bound," Prometheus is confident that his suffering will eventually end because he has knowledge of the future that Zeus needs to save his reign. In the sequel, "Prometheus Unbound," which survives only in fragments, Zeus apparently allows Heracles to free Prometheus. In this larger context perhaps Aeschylus must be said to have dimly grasped that Graph D (Section II, *supra*) best represents the relationship between technological growth and human well-being.

A brief reading of the Prometheus myth, and some others, in which the positive aspects are emphasized, can be found in the closing pages of David Landes, *The Unbound Prometheus: Technological Change and Industrial Development in Western Europe from 1750 to the Present*, Cambridge 1969.

[10] In *Works and Days* (106–201), Hesiod describes a series of Ages, beginning with the Golden Age. Though this was idyllic, it is not clear whether it was pre-technological.

[11] There are other myths in which man suffers because of *hubris* brought on by technology. See, for example, the flight and death of Icarus (Ovid, *Metamorphoses* VIII 183–235), and the death of Patroclus due to his misuse of Achilles' armor (Homer, *Iliad* XVI).

[12] In writing the appendix I have benefited (perhaps insufficiently) from conversation with Thomas Carpenter and Nicholas Smith.

ALBERT BORGMANN

TECHNOLOGY AND THE CRISIS OF LIBERALISM: REFLECTIONS ON MICHAEL J. SANDEL'S WORK

INTRODUCTION

This essay has a disciplinary and a substantive concern. The former is to promote the conversation between the philosophy of technology and mainstream political philosophy. These two disciplines too often ignore one another. Philosophers of technology should seek this conversation to secure wider consideration of the fruitful and consequential issues that will otherwise remain esoteric to their specialty. If philosophers of technology are unable to understand and overcome the disdain or fear of technology that they encounter in the priesthood of political theory, they can hardly hope to reach the lay people, the ones who finally matter.

Political and social philosophers for their part will remain caught in abstract, inconclusive, or inconsequential battles until they are willing to consider the daily and concrete circumstances that express and constrain our political aspirations. Those circumstances, whatever else they may be, are profoundly technological. The technological constraints on our political vision constitute the substantive concern of this paper. The resurgence of communitarian thought, I believe, is challenging the grip that technology has on our lives. Liberals seem to agree but fear that, once freed, we are likely to fall into more pernicious hands than those that presently hold us.

The issue is rarely put in these terms. It is transacted in charges of moral vacuity and impoverishment from the communitarian side and countercharges of vagueness and intolerance from the liberal camp. My concern is that this lively and promising debate will come to a pointless standoff unless it is rendered more concrete and fruitful through the consideration of technology. That the present controversy between communitarians and liberals harbors a deep-seated and disturbing issue is evident from the broad and sometimes passionate response that so well-tempered a treatise as Sandel's book on *Liberalism and the Limits of Justice* has received.[1] Michael J. Sandel was unknown when he wrote the book; it is an outwardly unassuming and moderately short piece.

105

Edmund F. Byrne and Joseph C. Pitt (eds.), *Technological Transformation: Contextual and Conceptual Implications*, 105–122.
© 1989 Kluwer Academic Publishers.

And Sandel has published little since. Why the extraordinary liberal response?

Sandel, I believe, has been able to call into question the adequacy or sufficiency of the moral order that many political philosophers have come to value and to cling to, John Rawls's theory of justice. At the same time, it seems to me, Sandel has not fully exposed the liabilities of our common order. But his work and the response it has received provide a fruitful opening for further exploration. Accordingly, I propose (1) to sketch briefly Sandel's critique of Rawlsian liberalism, (2) to survey and discuss the response to Sandel, and (3) to show that technology is the unspoken but crucial issue in the controversy between Sandel and his critics.

(1) SANDEL'S CRITIQUE OF LIBERALISM

For the sake of this essay's argument, I will take it that the Rawlsian or deontological version of liberalism is the liberals' best and most widely cherished hope. What then is Sandel's objection? In *Liberalism and the Limits of Justice*, Sandel sees the notion of the autonomous self at the center of Rawlsian or deontological liberalism. To say that the self is autonomous is to attempt to isolate what is morally crucial about human beings and what not. And expectedly, it is much easier to say what a human being, considered purely as a moral and autonomous person, is not.

The autonomous self must be free of accidental determinations. In its fundamental moral capacity and relations it must never be susceptible to such factors as its race, sex, or material possessions. These prohibitions are clearly in keeping with standard liberal democratic views. Indeed they represent liberal positions, now so widely shared, that they are embraced by any serious contender for national political power. But if wealth or race cannot bestow legitimate privileges in a democracy, how about a person's religion or cultural heritage? Can the latter serve as guides for the shaping of our basic social structures? Evidently, religion and culture are no more intrinsic to the autonomous self as such than sex or physical strength, and hence no particular religious or cultural concern can be embodied in the social framework itself though surely a person should be free to choose whatever religion within the fundamental social arrangement.

When it comes to religious and cultural values, the liberal position is no longer uncontroversial. Old money conservatives whose identity is bound up with a genteel upbringing and an august past could hardly recognize themselves or their selves in the antiseptically pure liberal self. Yet even new money conservatives whose fame and fortune rests not on inherited wealth or values but sprang from their native and intrinsic talents would not be secure in their self-understanding, given the force of the liberal argument. Yet if sex and race fail to be human qualities that are relevant to social morality, why should inborn intelligence or diligence be valued from the ethical point of view? Such genetic endowments, Rawls holds, are distributed by a natural lottery, i.e., in an arbitrary and morally indifferent way.

It is evident from the words I have used that there is a variety of relations that hold between the self and its natural and social endowments. There is no need to distinguish and explicate these different ties. What is crucial is that none of them matters in the determination of social justice. To grant any one a voice in moral matters is to subject the self to a morally arbitrary, i.e., heteronomous, power. The truly autonomous, i.e., self-determining self, relying only on itself, must be pure.

So far I have simply paraphrased Sandel's view of what is central to deontological liberalism, and it is certainly not obvious why this central notion of the self should turn out to be fatally flawed as the fundament of liberal justice. Sandel, however, goes on to derive two charges from his picture of the autonomous self. First and strictly speaking, it fails to yield in any rigorous way a theory of justice. Second, if we make certain concessions to the deontological enterprise so that its construction of justice gets underway, the resulting common order is bleak. Let me consider these criticisms in turn.

First, the autonomous self is unable to produce a vision of justice because its purity is indistinguishable from vacuity. Having been cleansed of all empirical and substantive characteristics, the pure self is devoid of resources to draw on. With Kant we may ascribe to it or grant it practical efficacy, i.e., we may think of it as capable of willing. It is then able to make choices. But as long as they are pure choices, unconstrained by anything, they will also be free of any guidance by principles. Hence they will be unprincipled or arbitrary choices. In particular, the parties to Rawls's original position, the selves that make the original choice of principles of justice, can have no reasons to agree on Rawls's two principles of justice.

Rawls, as one would expect, is aware of the deficiency of a purely autonomous self and has therefore enriched the austere Kantian position through the specification of the empirical circumstances that a theory of justice must acknowledge and is able to draw on. The view of these circumstances is inspired by Hume who saw in the conjunction of natural scarcity and human selfishness the need for an orderly social arrangement, i.e., for a theory of justice. For Hume justice would not be needed if nature were more bountiful or humans more generous. Justice is second best for him. But not for Rawls and the liberals. The austerity of the empirical and moral landscape that we behold in the circumstances of justice agrees well with the sublime severity of the autonomous self. Justice, therefore, is not just the remedy for the want of benevolence and abundance but the unsurpassable reflection of our status as autonomous moral agents. Justice becomes the first virtue simply.

The resulting order of liberal justice, however, is unacceptably cold and spare as far as Sandel is concerned. We can partake in its foundation only if we allow ourselves to be stripped of what finally animates us. As Sandel reminds us:

But we cannot regard ourselves as independent in this way without great cost to those loyalties and convictions whose moral force consists partly in the fact that living by them is inseparable from understanding ourselves as the particular persons we are – as members of this family or community or nation or people, as bearers of this history, as sons and daughters of that revolution, as citizens of this republic.[2]

In his book, Sandel was concerned to show the liberals that their recently favored moorings do not hold. In his more recent writings he argues that their vessel is dangerously adrift. This he does by outlining the chief structural features of the liberally conceived society and by showing that these are features of an unhealthy and unstable social arrangement. The relevant social configuration is captured by the title of Sandel's essay, "The Procedural Republic and the Unencumbered Self."[3] The two terms that are conjoined here represent a promise of liberty that remains unfulfilled and issues in the diremption of the fabric of our common order. The epithet "unencumbered" conveys the notion and appeal of liberation-from that is contained in the idea of the autonomous self, freedom "from the dictates of nature and the sanction of social roles."[4]

But what is wrong with the conjunction of the unencumbered self and the procedural republic? The answer is that it leads to a moral and structural decline. It leads to moral decline in that it erodes the democratic practices of the republic and allows power to drain away to judicial and bureaucratic institutions. It paralyzes us in our endeavors to restrain what we detest and to advance what we cherish. And it even leaves tolerance, the proudest of the liberal virtues, in jeopardy since the fundament of toleration lies in settled and confidently held convictions, not in "the confusion of atomized, dislocated, frustrated selves."[5]

This last point already suggests the structural decline that Sandel believes will overtake a common order based on unencumberedness and pure procedure. Such an arrangement first of all leaves people deeply dissatisfied. As unencumbered selves they feel detached and dispersed. Vis-à-vis the powerful machinery of the procedural state they feel powerless and trapped. And second, although the procedural republic overwhelms the individuals, it is not really stable and competent in its own sphere. It would collapse without the sense of community that it is in the process of etching away. And it is increasingly unable to cope with its economic, social, and diplomatic tasks.

What are we to do in the face of the moral and structural decline of the common order? Sandel does not advocate a turn to conservatism, old or new. Rather he appears to embrace a communitarian position that is distinctive over against both the liberal and the conservative parties. Its policies will sometimes overlap with liberal ones, at other times with those of the conservatives; and it has goals of its own. In each case, the policy flows from a consistent and idiosyncratic vision of the common order. At the center of that vision is the restoration of community.

(2) THE RESPONSE TO SANDEL

The response to Sandel has been of four kinds. (1) There are, not surprisingly, scholars of communitarian sympathies, Michael Walzer and Charles Taylor, who have welcomed Sandel's critique as a liberating endeavor, one that they take to have cleared a space in the liberal establishment for a communitarian alternative.[6] These scholars are joined by conservatives or centrists such as Charles Fried and Norman Care.[7]

(2) Two respondents, Richard Rorty and Jeffrey Stout, have praised Sandel from a postfoundationalist position though it is not clear that

Sandel would feel appreciated.[8] He may have shown that the liberal foundation is too thin to support and nourish vigorous communal life. But he nowhere says it is a mistake to search for a foundation. In fact, his argument is implicitly foundationalist. To point out that the foundation of a certain structure is weak and to imply that the structure is therefore unsound and liable to collapse makes sense only if structures are thought to rest on foundations. It is more doubtful still that Sandel would have much sympathy for the mildly hung-over postmodern bourgeois liberalism that Rorty recommends as the successor of foundationalist liberalism.[9] Stout, on the other hand, inclines toward the rich and venturesome political philosophy one may hope to see forthcoming from Sandel.

(3) A few critics, some of them chiefly located in one of the preceding or the following sort of response, take Sandel to task for failing to consider noncommunitarian alternatives to liberalism. Charles Fried, Amy Gutmann, and John Haldane have urged this point.[10] But their outlines of such alternatives are too sketchy or abstract to command much interest.

(4) Finally we come to a fairly strong group of mainstream liberal critics whose replies to Sandel are surprisingly unified. This group includes Lloyd Thomas, Brian Barry, Charles Larmore, Edwin Baker, Amy Gutmann, Jan Narveson, and Rawls himself.[11] They are more or less agreed in making two responses to Sandel.

The first amounts to a concession to Sandel. It jettisons the foundation of the autonomous self as unnecessary and claims that the twin pillars of deontological liberalism will yet remain firmly in place, one being the lexically primary commitment to equal rights and liberties, the other being the securing of a maximally value neutral common order. Sandel is charged with asking for too much. "Only" becomes a leading term in defining what is possible and needed.

The amazing fact is that in philosophy and politics the liberals have become the conservatives.[12] They are no longer the bringers of liberating news but now see themselves as occupying the high ground and as the defenders of the venerable heritage of the Enlightenment. A celebrated instance of this defensiveness is Habermas's protracted struggle to recover and uphold the critical and liberating force of modernity.[13] An interesting sidelight was the 1987 Baccalaureate Address of Benno Schmidt, President of Yale, whose charge to the graduate class was this:

We charge you to seach for meaning and value in your lives by standards of intelligence, and to carry these standards into a world that inclines to immediacy, subjectivity, and cynical incompleteness.[14]

The fervor of the defense is invariably fueled by the conviction that no alternative to the liberal arrangement can be trusted. Mistrust, at times, shades over into hostility, and Sandel has received his share of it. It was most strongly expressed by Bryan Barry who said:

Sandel makes the transcendence of justice by group identity sound very high-minded. But it gives the green light to every string-pulling parent and crony-hiring academic. And at the end of that road stand Torquemada, Stalin, Hitler, and Begin. Sandel's argument should be turned on its head: it is exactly when "devotion to city or nation, to party or cause" run deepest that the constraints of justice on the pursuit of those allegiances are most needed.[15]

I do not think that the liberal accomplishments of equality and tolerance, such as they are, will be undermined and collapse. To be sure, they need and deserve everyone's vigilance. The crucial plight of the liberals, however, is the stagnation of their program at a point that is so very far from their goal. The order of inequality in this country has hardened in the last two decades. The force of the rhetoric of autonomy and equality is spent.

The predicament of the *philosophical* liberals is most evident from the programs and arguments of the *political* liberals. None of the latter any longer attempts to promote a liberal program through promises of greater liberty and equality. All of them appear to think that, if there is to be political movement, it can only come through technocracy, through the advancement of technology and a rise in prosperity. Greater social justice, so it is hoped, will come in the train of technological progress.

This is a distressing practical setback for the renaissance of liberal theory that had begun with Rawls's great book. The prior utilitarian liberalism had been objectionable not because its practitioners were indifferent to social justice but because they had entrusted it to a vehicle, utilitarianism, that failed to guarantee safe passage for the inviolability of individual liberty and dignity. Rawls received such an enthusiastic reception because his theory assured the safeguarding of the liberals' core conviction within an admirably resourceful and circumspect social design. But the newly found theoretical confidence of the liberals was weakly matched by the liberal policies of the seventies and appears to leave the liberal practitioners of the eighties distrustful and unmoved.

The liberal theorists need to ask what sense we are to make of the displacement of their program by a trust in technology as the wellspring of social progress. Sandel faces the same task from a different angle. He can carry his critique of liberalism forward only if he is able to show that liberal democracy without foundations is at the mercy of technocracy and can therefore not afford to be in different to the ambitious issues he has raised. And he needs to confront technology more directly to avoid for his part the ground swell of technological fixes and to claim firm ground for the recovery and realization of community.

(3) THE PROBLEM OF TECHNOLOGY

I have been using "technology" broadly so far. Rather than providing a formal and detailed definition at the start, let me begin by circumscribing technology broadly as the form of life that is characteristic of an advanced industrial society. Technology, always short for "modern technology," is a distinctive way of taking up with reality and results in characteristic artifacts, large and small. It is a unique convergence and configuration of scientific, economic, and cultural factors.

To give this outline detail, I begin by showing to what extent Sandel's views are consonant with the decisive features of the technological society. I will then proceed to points of divergence between Sandel's critique and the pattern of technology. This will lead us to a deeper level of the analysis of technology. At that level, I believe, we can, if not adjudicate, at least illuminate more hopefully what divides Sandel and his liberal critics.

Sandel's concept of the unencumbered self and the liberating vision it holds out are at the origin and center of the character of modern life. The promise of liberty arises out of the founding liberation movement of modernity, viz., the Enlightenment. The latter is commonly taken to be an intellectual, political, and abstractly cultural movement, a view that overlooks the immediate and pervasive ways in which this promise was enacted through or as technology. Today the rhetoric of exhilarating freedom is found most often and prominently in advertising. Look at the advertisements in, say, the *New York Times Magazine*, and you will see a celebration of the self, encouraged and empowered to shed all bonds and burdens and to reconstitute itself as a subject of possession, i.e., as a self that defines itself through the commodities it chooses to possess. But any such possession remains at a distance to the self. One is free to discard the old and acquire the new, yet the new itself is offered on the understanding that it will leave us free and unencumbered. The self has no abiding traits or attachments. It is fashionable but will not become old fashioned. Its defining quality is its unqualified freedom to choose. We are reminded of the title of Robert Musil's great novel, *The Man Without Qualities*.

But has it not been said that the strictly autonomous self, pure of all constitutive properties and attachments, is devoid of direction and sustenance? How in technology does the unencumbered self escape vacuity or dissolution? In theory, the circumstances of justice are invoked to provide a context for the self, and within this setting, the machinery of the procedural republic procures the structure and stability needed to support and define the self. In practice the technological machinery draws on the conventions of human shapes and movements, on the heritage of gentility and luxury, and on traditional notions of beauty and excellence as raw materials to procure commodiously available possessions for the self.

However, no one is really given to any illusions about the narrow constraints on the discretion of the unencumbered self. Unencumberedness is indulged in leisure and consumption. On Monday morning it is back to work. Everyone understands that one must do one's part in sustaining and expanding through labor and production the vast substructure that supports the glamorous surface of life. Submission to the discipline of labor and production is often grudging but acknowledged as inescapable just the same. At any rate labor contrasts sharply with the sense of disburdenment that we feel in the sphere of leisure. This dividedness of life is reflected in Sandel's observation that "the liberal self is left to lurch between detachment on the one hand, and entanglement on the other."[16]

Of course, Sandel is not considering technology directly. As a term, he mentions it only in asides and obliquely. "Except on the bleakest of technocratic visions," he says in one place, "political questions are inescapable moral questions . . ."[17] And in noting the grave political consequences of the events around the turn of the century when "national markets and large-scale enterprise displaced a decentral-ized economy," he refers to a development that was fueled and guided by modern technology.[18]

What is crucial for the argument of my essay is not the question whether Sandel gives a particular vocable, technology, terminological prominence or not but whether he is equal to a conception of political philosophy that he defines very well. "This is the sense," he says, "in which philosophy inhabits the world from the start; our practices and institutions are embodiments of theory." Hence the salient question: "What is the political philosophy implicit in our practices and institutions?"[19]

Sandel's distinction between the procedural republic and the unen-cumbered self captures the central feature of the technological order, a division and correlation of regions that, as suggested, are well instan-tiated by the daily detail of our technological practices and institutions. It is a pattern that is further reflected in the distinctions and correlations of the public and the private, of production and consumption, of machineries and commodities, of means and ends. Sandel is also correct in seeing our relation to this order as uneasy and divided, as a lurching between detachment and entanglement.

I want to suggest, however, that Sandel's position can be clarified and strengthened if on the one side we stop looking for aid from quarters whence it is not really forthcoming and on the other side discover and claim support that is commonly invisible, yet potentially powerful. As was apparent above in section one, Sandel ties his plea for a renewal of community to the implicit warning that without it our society will suffer a structural decline if not collapse. But I fear that if this warning is to turn people toward community, they will remain unmoved, sensing rightly that the technological order is at bottom more resilient and self-righting than appears at the surface.

At the same time this common order is more discriminatory and intolerant than we commonly realize. The reprehension, implicit in this claim, is not *that* technology discriminates, but *how* it does so. In fact it seems obvious to me that any common order, inasmuch as it orders our common affairs, inevitably channels individual undertakings into one direction rather than another.

If it is possible to show that and how the order of technology banks and directs the social currents, the dispute between Sandel and his critics moves to a new level where the controversy is no longer between communitarian constraints and liberal freedom but between communitarian discrimination and technologically specified liberal discrimination. At this level it can no longer be taken for granted that communitarianism is the defendant. Technological constraints may be more debilitating and less favorable to the good life than communitarian ones.

All I can do within present limits to advance this claim is to provide suggestive illustrations. The first illustration is meant to reveal the resilience of the technological order and the allegiance it commands. The illustration comes from developments in the insurance industry over the past few decades. The insurance system in this country is in a crisis at least in part because increasingly strict construals of liability and ever broader construals of harm have been argued by lawyers and accepted by the courts. Consequently more and larger awards have been granted to plaintiffs to the point where liability insurance is becoming prohibitively expensive or altogether unavailable.[20]

What is at the bottom of these particular changes? Consider the owner of a lawn mower who pulls the starter cord until he suffers a heart attack. Whose fault is it? Not, ultimately, the owner's according to the doctrine of strict liability. Or consider someone who sees a close relative crushed to death in an accident. Who is to bear the burden of this trauma? Not the victim's next of kin who will be compensated for mental suffering and grief. We are no longer willing, it seems to me, to take responsibility for certain actions that are clearly ours nor for those trying events that, in Sandel's memorable words, "I neither summon nor command." But note the sense of the whole sentence from which this phrase is taken. "For to have character," Sandel says, "is to know that I move in a history I neither summon nor command, which carries consequences nonetheless for my choices and conduct ." What Sandel imputes to the deontological liberal theory, viz., that it is led to imagine the self as "a person wholly without character, without moral depth" is actually coming to pass.[21] In our desire to disburden ourselves ever more radically we are becoming characterless and morally shallow. We are increasingly unwilling to assume responsibility not only in regard to accidents, but also vis-à-vis parents, spouses, children, and the frailty of our bodies. This is evident, I believe, from the fate of old people, from divorce rates, from the declining well-being of adolescents, and from the senseless anger with which we often greet sickness and physical debility .

It might be objected, however, that these developments constitute a shifting of responsibility from one class of persons to another rather than a general refusal and dissipation of responsibility, viz., shifts from clients to lawyers, from patients to physicians, from consumers to producers; and the latter, one might say, are as straightforwardly responsible as humans have ever been. But the grief over the loss of a loved one does not become the occasion for an attempt to win a million dollars without the consent and cooperation of the grieving person. And plaintiffs in suing a doctor or producer know that they are not imposing a crushing punishment on a real individual person. Rather they know that they are seeking the relief of money through a person from the impersonal machinery of the insurance industry. In fact, few lawyers would take on a case where the defendant would not be the conduit to an insurance company but one whose very substance and well-being would be destroyed if found guilty.

And who or what, finally, enables the insurance industry to make these huge payments to successful plaintiffs? We all do as rate payers, and we are willing to do so in line with the broad and implicit compact according to which individual disburdenment will be paid for through support of the common technological machinery. To be sure, the machinery has its requirements of efficiency, and these may conflict with individual demands. But such a conflict is ultimately a technical problem, and it will be solved as such. It will not be the occasion for moral reflection and reform.

This phenomenon is well reflected in the jurisprudential arguments supporting strict liability. The advocates are at pains to reduce or exclude the problem of culpability. Strict liability is liability "without fault."[22] It is to serve not as a measure of guilt, and when ascertained, as an expression of condemnation but as a device for the prevention of an extensive social harm. Accordingly, the judicial response to a finding of strict liability is not meant to be the punishment of a crime but the assessment of a penalty for a technical violation.[23]

Christine Sistare fears that the adoption of liability is "inefficient and self-defeating."[24] And in March of 1986, *Time* warned that America was "well on the way to insuring itself to a silly, shuddering halt."[25] But we have continued to move along fairly well. Why? Not because we have recovered a sense of civic and moral responsibility but because we have recognized the need for technical fixes of the judicial machinery and have undertaken repairs sufficient to keep the machine functioning.

And yet Sandel's intuition that there is something profoundly debilitating in this arrangement is correct. But what Sandel has not shown so far but what an analysis of technology, I believe, strongly suggests, is this: The profound debility of the present social arrangement is distressing on the liberals' own terms. The argument, briefly, is that the common technological order is in its way as constraining and discriminating as a community. Or in other words, the technological order fails to meet the liberal ideal of social openness no less than does a community. If this is true, liberals can no longer reject communitarianism in the name of a morally superior liberal order.

But is it true? To answer, let me begin by distinguishing two kinds of discrimination that attach to the technological order. The first kind is visible within the established order. I will call it intraordinary or simply ordinary discrimination. It consists in the unequal distribution of the benefits as they are understood within the technological order. Liberals, unlike conservatives, are keenly sensitive to them and concerned to reduce them. But such reforms are possible within the established order of the unencumbered self and the procedural republic. The liberals would express this point by saying that what we need is not a radically new common order but rather the realization and fulfillment of the best aspirations of the present order. This view in turn finds expression in the defensiveness with which liberal critics have met Sandel's critique. The present liberal-technological order is also liable to extraordinary discrimination. It condemns certain legitimate aspirations and their holders to a marginal existence, to a diaspora, and even to a sort of exile. The present common order is unable to invite these people in and to be generous to their deepest concerns unless it changes itself in a radical and extraordinary way.

Before I attempt to illustrate this claim, I should clarify a point that has been implicit so far. It concerns the tie between technology and liberal democracy. How close and necessary is it? The consonance of Sandel's analysis of liberalism with the technological character of our lives and the emergence of analogous correspondences throughout the preceding reflections indicate a crucial bond between our political and material order. Liberalism, in fact, would have remained utopian and unrealizable had it not been specified technologically. Of course liberalism is not tied to technology in every sense of the latter word. The point is that the liberal democratic order requires a very specific material culture, and technology, I believe, is an illuminating title for that culture.

Turning then to the extraordinary discriminations of the technological culture, how do we come to feel their force? Technology as a way of life is now so deeply entrenched that its basic pattern has become all but invisible, and the constraining force of the pattern is often experienced in ways that we take to be accidents of normal frustration or indications of personal failure. One event, however, often elevates us above our implication in the established order and allows us to see it in a clearer way. It is the event of becoming a parent, and it will serve as my second suggestive illustration. "The great conservatizing experience is having children," says George F. Will.[26] While it may not make us into conservatives, it does sharpen our sight for the liabilities of the liberal-technological order. Even so insistently a conventional economist as Robert J. Samuelson, the *Newsweek* columnist who reacted with anger and derision to the Catholic bishops' violation of economic orthodoxy, spoke with moving candor about the limits of mainstream economic wisdom when he contemplated it as a father.[27]

What do we learn about the present order as we raise our children? We see them grow up with a mixture of fear and wonder; wonder at the vigorous spontaneity with which they unfold their capacities and appropriate the world; fear about our ability to provide them with the setting that would be equal in wealth and depth to their talents. We reflect on the limitations and deprivations that we have suffered as children and youngsters and resolve to provide our children with the instruction and surroundings that will give them the intellectual skills, the physical valor, the artistic competence, and the habits of compassion that we failed to attain as fully as we should have.

This endeavor succeeds well enough as long as our children are fully within our own care. Kindergarten and grade school will breach the walls of excellence and allow waves of gimmickry, distraction, and competitive violence to rush in. But until puberty the parents are able to be the beacon and haven of the child's world. Yet it is a secluded and thereby artificial world that we have created. This we realize painfully when through adolescence and in high school our children begin to let go of us and we must learn to let go of them.

I am speaking here of the normal child, not of the genius whose extraordinary gifts rush him along past all perils to a splendid goal of excellence, nor of the maverick whose fierce independence repels her, for better or worse, from the blandishments of conformity. Normal children will want to enter the world at large and join with people of their age. A family is then a community of three or four, contending with a society of hundreds, ·thousands, and millions. The power and status of this society is so massive and unquestioned that parental opposition to it will inevitably appear as capricious authoritarianism.

But why would you as parents want to oppose society? Neoconservatively inclined critics of the liberal society have pointed out that it is an increasingly dangerous world where the rates of juvenile delinquency, drug and alcohol use, unwanted pregnancies, and death are rising.[28] No one would want to belittle the heartbreaking consequences of these developments. Still, their terror and pain are clearly visible within the established order. In that sense the overt dangers to adolescents are ordinary, and they may be solvable in ordinary ways, i.e., through technological fixes. At any rate, the strictly extraordinary harms and discriminations are to be found where juvenile matters seem safe and harmless. The area of the contemporary culture to which you are forced to surrender your children is that of leisure and consumption, of cars, videos, MTV, cheap movies, and fast foods. The problem with this milieu is not that your child fails to pursue excellence in it but that it simply excludes the possibility of excellence in a traditional sense. Excellence is displaced by its bastard sibling affluence. More money allows a youngster to have a more glamorous car, finer foods, and a better stereo system. But these possessions are as accidentally attached to a person as talents are to the unencumbered self. It is of no help to insist on connecting adolescents with the totality of the present culture, i.e., to make them labor as well as consume. Cooking the hamburgers at McDonald's or selling the stereos at Sears' is as mindless and as little conducive to excellence as the consumption of commodities.[29]

Assume, none the less, that your son or daughter has not succumbed to drugs, has not been killed in a car accident, and has not become an unwilling parent. Assume even more, viz., that having experienced the deeper pleasures of excellence your son or daughter has tired of consumption and become devoted to the pursuit of an engaging and ennobling life. She or he is admitted to an excellent university, and there is now an episode of four years where this youngster revels and is schooled in some of the finest accomplishments of body and mind that have come down to us over thousands of years. In other words, your child acquires what we fondly and proudly call a good liberal arts education.

These are four exhilarating and idyllic years that recall at the highest level the wonders and promises of childhood. But like childhood, the higher liberal arts education takes place in a cultural enclave. This will become obvious when your child has received the baccalaureate degree. What now? It is not that your son or daughter fails to have goals or plans. He or she may want to become a writer, a musician, an artisan, or a farmer. Let us assume that we know that your child would be competent at any of these callings, not a genius, but one who would do valuable work. But when it comes to the professions just mentioned, it is all or nearly nothing. You have to be a genius or lead a marginal existence, a life without economic comfort and devoid of the interest and approval of the world at large.

My point in these reflections has been to suggest that there is a substantial and legitimate, even admirable, way of life that is unable to prosper within the established order. To make the point more persuasive we would have to generalize and explicate the way of life and also bring out in sharper outline the crucial features of the inhospitable established order.[30] This I can do here in only the most summary way. To begin with the present order, it divides our lives sharply into the areas of leisure and labor. The availability of the greatest number and variety of consumable commodities is the avowed standard of living. Labor and production are subordinated to this avowed goal and constantly transformed entirely in accord with it. Our common efforts are devoted to the security and improvement of consumption and production and to a small extent to increasing the equality in the distribution of the privileges of leisure and labor. This social order is hostile to a way of life which is dedicated not to consumption but to the excellence of focal things and communal practices, not to labor regarded merely as a means for the safety and advancement of consumption but to work as an intrinsically ennobling engagement with the material world and fellow humans.

If the discussion could be moved to this level of exploration and discourse, the communitarians might be able to engage the liberals more resourcefully. The liberals I know well are sensitive to the communitarians' concerns and often accomplished in them. But, to join Williams's and Rawls's terms, the liberals tend to convert their thick ethical convictions into the thin currency of liberal justice and ultimately back again into the thick social overlay of technology.[31]

When liberals are disturbed by the violation of the sacredness of nature, they call for tighter zoning, soil conservation, and the multiple use of forests. When they are distressed by the degradation of the quality and dignity of work, they inveigh against multinational corporations or push for better funding of OSHA. When they witness the demise of communal celebration, they are either silent or work against their intuitions by shoring up every breach they can discover in the wall that separates church and state.

More generally, the liberals convert the topics of worship, family, tradition, care, and excellence into issues of social justice and environmental and international security. And what they actually secure and share more fully are, by default, the apparently sleek and antiseptic commodities of technology. Liberals are embarrassed to bring up in public the thick and awkward matters that sustain them privately. But if they came to see that technology has its own distressing thickness, one that overlies and suffocates our best aspirations, they might join Sandel in exploring the ways that would lead us to a more engaging and communal life.

NOTES

[1] Michael J. Sandel, *Liberalism and the Limits of Justice* (Cambridge: Cambridge UP, 1982).

[2] Sandel, *Liberalism* 179.

[3] Sandel, "The Procedural Republic and the Unencumbered Self," *Political Theory* 12(1984): 81–96.

[4] Sandel, "The Procedural Republic" 87.

[5] Sandel, "Morality and the Liberal Ideal," *New Republic* 7 May 1984: 17.

[6] Michael Walzer, "From Contract to Community," *New Republic* 13 Dec. 1982: 35–37; Charles Taylor, "The Nature and Scope of Distributive Justice," *Justice and Equality Here and Now*, ed. Frank S. Lucash (Ithaca: Cornell UP, 1986) 47–48.

[7] Charles Fried, "Liberalism, Community, and the Objectivity of Values," *Harvard Law Review* 96 (1983): 960–68; Norman S. Care, rev. of *Liberalism* in *Nous* 19(1985): 459–67.

[8] Richard Rorty, "Postmodern Bourgeois Liberalism," *Journal of Philosophy* 80(1983): 586; Jeffrey Stout, "Liberal Society and the Languages of Morals," *Soundings* 69(1986): 32–59.

[9] See also Rorty, "Method, Social Science, and Social Hope," *Consequences of Pragmatism* (Minneapolis: U of Minnesota P, 1982) 191–210.

[10] For Fried, see note 7 above. Amy Gutmann, "Communitarian Critics of Liberalism," *Philosophy and Public Affairs* 14(1985): 320–22; John J. Haldane, "Individuals and the Theory of Justice," *Ratio* 27(1985): 189–96.

[11] D.A. Lloyd Thomas, rev. of *Liberalism*, *Philosophical Books* 24(1983): 249–51; Charles Larmore, rev. of *Liberalism*, *Journal of Philosophy* 81(1984): 336–43; C. Edwin Baker, "Sandel on Rawls," *University of Pennsylvania Law Review* 133(1985): 895–928; Amy Gutmann, "Communitarian Critics," 308–22; Jan Narveson, rev. of *Liberalism*, *Canadian Journal of Philosophy* 17(1987): 227–34; John Rawls, "Justice as Fairness:

122   ALBERT BORGMANN

Political Not Metaphysical," *Philosophy and Public Affairs* 14(1985): 238–39.

[12] Cf. Michael W. McConnell: "Now it is the left that cherishes stasis and precedent, that is fighting a rearguard action against change." "The Counter-Revolution in Legal Thought," *Policy Review* no. 41 (Summer 1987): 19. I agree with McConnell, but I do not share his glee.

[13] See Stephen Watson, "Jürgen Habermas and Jean-François Lyotard: Post-modernism and the Crisis of Rationality," *Philosophy and Social Criticism* 10(1984): 1–24; Wolfgang Welsch, "Vielheit ohne Einheit?" *Philosophisches Jahrbuch* 94(1987): 111–41.

[14] Benno Schmidt, "Commencement," *Yale Alumni Magazine* 50, no. 8 (Summer 1987): 41.

[15] Barry 525.

[16] Sandel, "The Procedural Republic" 91.

[17] Sandel, "The State and the Soul," *New Republic* 10 June 1985: 38.

[18] Sandel, "The Procedural Republic" 92.

[19] Sandel, "The Procedural Republic" 81.

[20] "Sorry, Your Policy is Cancelled," *Time* 24 Mar. 1986: 16–26.

[21] Sandel, "The Procedural Republic" 90.

[22] Christine T. Sistare, "On the Use of Strict Liability in the Criminal Law," *Canadian Journal of Philosophy* 17(1987): 395–96.

[23] Sistare 398–405.

[24] Sistare 407.

[25] *Time* 16.

[26] George F. Will, *Statecraft as Soulcraft* (New York: Simon, 1983) 78.

[27] Robert J. Samuelson, "The Bishops and Adam Smith," *Newsweek* 3 Dec. 1984: 75 and "In Praise of Children," *Newsweek* 13 Jan. 1986: 49.

[28] Peter Uhlenberg and David Eggenbeen, "The Declining Well-being of American Adolescents," *Public Interest* 82(1986): 25–38.

[29] Ellen Greenberger and Laurence Steinberg, *When Teenagers Work* (New York: Basic, 1986).

[30] I have tried to do that in *Technology and the Character of Contemporary Life* (Chicago: U of Chicago P, 1984).

[31] John Rawls, *A Theory of Justice* (Cambridge, MA: Harvard UP, 1971) 395–99; Bernard Williams, *Ethics and the Limits of Philosophy* (Cambridge, MA: Harvard UP, 1985) 129, 143–45, 200.

CLIFFORD CHRISTIANS

A THEORY OF NORMATIVE TECHNOLOGY

I argue in this paper against the commonplace that technology is neutral
– that only the uses involve human valuing. And given the value-
ladenness of technology, I suggest a normative framework by which the
technological process – design, fabrication, and use – can proceed
responsibly.

TECHNOLOGY'S SUPPOSED NEUTRALITY

We too readily assume that technology is merely a tool which can be
used rightly or wrongly. As Oxford's R.A. Buchanan has written,
"Technology is essentially amoral, a thing apart from values, an instrument which can be used for good or ill."[1]

In this view, if the cancer rate goes up because of indiscriminate use of
pesticides, if whales become extinct because of modern hunting
methods, if catastrophic climate changes occur because of too much
carbon dioxide in the atmosphere, and if new technologies devastate a
remote tribal society, we are not to blame the persons who designed and
manufactured the technological products involved. They are absolved of
responsibility. Instead we should blame the politicians, corporations
and individual consumers for their bad judgment. One is reminded of
the familiar slogan: "Guns don't kill people, people do." Technology is
not at fault, but the uses to which we put it.

In Arnold Pacey's example, in Swedish Lapland snowmobiles are
used for reindeer herding, among Canada's Eskimos for trapping and in
Wisconsin for leisure.[2] Technological products are independent; they
can be used to support completely different cultures and lifestyles. A
knife in the surgeon's hands saves a life and takes it away when used by
a murderer. The same projector shows pornography and National
Geographic Specials.

A professor from Beijing University said on our campus recently that
the Chinese believe technology is neutral. In fact, he insisted that this
belief is vital for China's future. Only on the basis of neutral technology

123

Edmund F. Byrne and Joseph C. Pitt (eds.), Technological Transformation:
Contextual and Conceptual Implications, 123–139.
© 1989 Kluwer Academic Publishers.

can government officials be persuaded to relax their control; only if technology is neutral can the Chinese ever expect competition among research institutions and companies.

As the philosopher George Grant reminds us already in the title of his penetrating essay, we are beguiled into believing that "(t)he computer does not impose on us the way it should be used."[3] "The computer does not impose" holds before us a worldview with neutral instruments on one side and human freedom to use and control them on the other.

Describing the rationale for this dominant view is beyond my purpose in this essay. Let me cite only three reasons briefly as background for understanding the importance of a normative theory.

1. We are often trapped by the moral philosophy which separates means and ends, with technology then viewed as a neutral means to reach a predetermined, value-laden goal.

2. In part, the neutrality position arises from the notion that *science* is natural, and since science is closely linked with modern technology, the neutrality of technology is assumed too in the popular mind.

3. Or one could argue that should and ought and moral imperatives have virtually disappeared from our vocabulary, ergo, out of this cultural phenomenon, by default, the technological enterprise is left to mechanical terms only.

DEFINITION OF TECHNOLOGY

Regardless of its origin, the prevailing opinion that technology is neutral typically focuses on hardware, on tools and mechanical artifacts. With Pacey, I find that definition deficient in scope and view technology instead as a human process value-laden throughout.[4] Technology in my view is the distinct cultural activity in which human beings form and transform neutral reality for practical ends with the aid of tools and procedures.[5] On this view, I would argue that valuing penetrates all technological activity, from the analytical framework used to understand technological issues, through the processes of design and fabrication, to the resulting tools and products. Although valuing is surely involved in the uses to which people put these technological objects, valuing saturates every phase prior to usage as well.

There is no isolated, neutral definition of problems, solutions, and concepts, all existing in a presuppositionless vacuum. Instead, technology proceeds out of our whole human experience and is directed by our ultimate commitments. Technology is value-laden, the product of the primordial valuing activities of human beings. Value-loading is embedded in technologies themselves. It does not only arise as technology interacts with political and social factors, but emerges from the basic fact that technology objects are unique, not universal.

Technological tools and products are particular. They combine specific resources – know-how, materials, and energy – into unique entities, with unique sets of properties and capabilities. Any technological object, therefore, embodies decisions to develop one kind of knowledge and not some others, to use certain resources and not others, to use energy in a certain form and quantity and not some other. There is no purely neutral or technical justification for all these decisions. Instead, they arise from conceptions of the world, themselves related to such issues as permissible uses, good stewardship, and justice.

Technological objects are also unique in that they do impose on us the way they can be utilized. There is frequently some degree of latitude as to how a technological object can function, but obviously no complete freedom. A nuclear bomb, for example, limits the ways in which it can be used in modern warfare. In general, each object is designed to function in a certain manner, utilizing particular kinds of inputs and producing particular outputs. Even a simple technological product such as a can opener must be used in a certain way if it is to fulfill its function of opening cans effectively.

Computers open up some choices and close down others, such as the sort of information that can not be handled by the system. Air travel makes certain options possible, but it closes others, such as schedule and destination flexibility, or the ability simply to stop and look at the scenery. The alteration in the pattern of available choice is not random, but depends on technology's structures.

Any technological artifact embodies particular values which by definition give to that artifact properties which other artifacts do not possess. Harold Innis and Marshall McLuhan recognized that fact regarding communication technologies, for example. Innis argued that each medium is biased toward space and time in particular ways – print biased toward space, oral communication toward time, television toward immediacy, and cinema toward visual realism.[6]

As the complexity of technological objects increases, the rigidity of the regimen which must be followed in their use normally increases. Thus George Grant argues: "Abstracting facts so that they may be stored as 'information' is achieved by classification, and it is the very nature of any classifying to homogenize what may be heterogeneous. Where classification rules, identities and differences can only appear in its terms." This means, he concludes, "that computers increase the tempo of the homogenizing process in society." This unique entity we call the "computer" embodies specific capabilities and restrictions in usage.[7] And these restrictions are anything but value-free, as documented in Joseph Weizenbaum's early *Computer Power and Human Reason* and Theodore Roszak's recent *Cult of Information*.

In short, every technological object is unique, with a particular set of properties embedded within it. Therefore, inherent to it are boundaries on the way it can be used, and these boundaries are composed of choices from available materials, information, and energy.

The technological process carries within it the values that people have inevitably – even if unconsciously – placed there. Obscuring this fact with presumed neutrality has been very costly. For one thing, it typically allows scientists and engineers to escape responsibility. And in its heaviest form, it promotes a version of technological determinism in which technology's own inner logic appears to drive its development. Moreover, it leads to an exaggerated, unbalanced emphasis on magnitude, control, uniformity, and integration – what Pacey calls the virtuosity values. The narrow view fosters the working rule that "If it can be done, it should be done," pushing out other significant dimensions in our decision-making.[8]

THE IMPORTANCE OF NORMATIVITY

But the claim that technology is a value-loaded enterprise is only a first-level approximation, which left undeveloped suffers from Mannheim's paradox:

That is, if truth is culture-bound, then within our own culture we ought not see our culture-bound truth as absolute. Thus we cannot advocate value relativism without rising above it and we cannot rise above it without giving it up.[9]

The only valid option, it seems to me, concerns our choice of the supremely valued, that is, the norms by which our value judgments are ordered. Otherwise I have only presented the innocuous argument that valuing is expressed preference and nothing more. Known as an emotive view of values, this perspective makes only the minimalist claim that to value something says I like it. In contrast, I would contend that the values which infuse the technological process must simultaneously be considered process transcendent.

These are debatable matters to be sure, but at least a general indication why the only legitimate alternative to neutral technology is normative technology. Human beings, peculiar among living organisms act upon a structural whole, a universum, rather than in terms of natural surroundings or immediate experience only. Therefore, a defensible version of value-ladenness ultimately depends on the ontological status of its normative claims.

Recovering the notion of normativity, I believe, is the catalyst for an authoritative philosophy of technology. Regardless of the particular philosophy one may consider most defensible, I invite readers to the theoretical trenches. What do norms entail epistemologically? What is their ontological status?

The question appears in countless forms: Are there universal truths? Have transcendental criteria or aprioris been emasculated since the Enlightenment and therefore unavailable for us in our theorizing? Note the conclusion of Hans Jonas:

The very same movement which put us in possession of the powers that now have to be regulated by norms — the movement of modern knowledge called science – has by a necessary complementary eroded the foundations from which norms are derived; it has destroyed the very idea of norm as such.[10]

Given this reality, one works the normativity domain as an act of conscience. The question is whether theorists of technology can restore to their agenda a normative foundation from which to evaluate, judge, and reconstruct all phases of the technological process.

Over the centuries, in Western scholarship at least, philosophers could take divine command theories seriously or, as a minimum, presume various versions of Platonic absolutes. Even as these presumptions unraveled in the 19th century, the immutability of nature, which took care of itself and demonstrated permanence of a physical kind, made norms of semi-enduring status at least imaginable. In other words, as Jonas makes clear, all previous ethics reckoned only with noncumulative phenomena, directly or in the background. Morality could conceivably be the property of all, living as they did on terra firma before the face of an Eternal being, or at least with a first principle that ordered the vacillations of everyday affairs. But as Jacques Ellul, Ivan Illich and others have demonstrated, a technological age is cumulative, expanding, augmentative. Today the technological world is designed and maintained by experts, and ethical imperatives cannot be constructed without taking such expertise as the pivotal moment.

In the face of intellectual and cultural shifts of this magnitude, the tendency is toward cynicism or nihilism or toward rejection of normative claims as moralistic cries in the wilderness. Or we retreat into a minimalist position that refuses to confront the major issues and reconceptualize the foundations. Or the stalwart do-gooders presume an evolutionary dialectics; that is, that violence and peace, bonum and malum, capitalists and socialists, modern and primitive will evolve into a world community as mutual understanding grows; what we need, on this view, is agreement on a tolerant process not precommitment to absolute truths.

Against this background, I believe a substantive advance in the philosophy of technology depends on our validating the idea of normativity as a necessary though insufficient condition. Theories of technology are an ideological exercise in the absence of normative criteria. Making norms intellectually unassailable is too ambitious for this essay, and perhaps impossible anywhere. My intention is to establish a culture of discourse in which normativity as a minimum becomes *herausforderend* in Heidegger's terms – pressing itself upon us with compelling force.[11] When students of communication technology, see normativity as *herausforderend* – even if all the metaphysical issues are not resolved – our research and writing will become of enduring significance.

CULTURAL CONTINUITY

The theory of normative technology I consider defensible is controlled by the norm of cultural continuity. Cultures I take to be those patterns of belief and behavior which orient life and provide it significance. They constitute the human kingdom by organizing reality and indicating what we ought to do and avoid. Cultures are humankind's distinctive and immediate environment built from the material order by men and women's creative effort. Cultures are those webs we stitch together to direct societal practice, the embodiments of our living as persons-in-community. Max Weber called them the "webs of significance" within which the human species uniquely lives. Echoing the same theme, anthropologists since E.B. Taylor have contended that we are dependent on the inheritance and transmission of cultural forms or we cannot exist as a species. I speak of a universal practice – all humans are cultural beings.

Therefore, to place this controlling norm in more technical language, note the formal criterion: Technological products are legitimate if and only if they maintain cultural continuity. This value-laden enterprise we call technology, to be set in a normative direction, must comport well with cultural continuity. We ought to consider the viability of historically and geographically constituted peoples to be nonnegotiable. Given culture's centrality in our humanness, its continuity warrants the primary focus.

I am suggesting that our technological activities must "comport well" with the controlling norm of cultural continuity. "Comport well" carries with it a creative ambiguity. On the one hand, it asks for a strong relationship, one rooted in cultural continuity. On the other hand, it does not specify this relationship precisely. This does not suggest that all technological activities are detailed in that first principle, nor that they can be deduced directly from it.

I thereby insist on humility about human knowledge. No one should expect to find a complete system of normative standards for technological activities that results by simple deduction. All human effort – including theorizing – is done by beings whose knowledge is incomplete, whose insights are imperfect, and whose understanding is often blinded by tentativeness.

And, of course, I speak of historical continuity. I am not proposing a static view of continuity without discontinuity; that, at best, represents an ancient Greek cyclical view of history. But I choose to emphasize continuity so as to contradict notions of blind evolutionary progress that undervalue continuity altogether. Also continuity undercuts the modernization schemes devised by transnational companies or colonial powers to strengthen their hegemony. Perhaps oscillation conveys the appropriate image of a dialectic between continuity and discontinuity – with the overall pattern one of cultural formation rather than rejection.

Clearly, opposites are at work here – differentiation and integration, centralization and decentralization, large scale and small scale, uniformity and pluriformity. And difficult choices must often be made between those opposites. But the prevailing direction is always toward differentials, decentralization, small scale, and pluriformity. Sometimes that results in a break with past practices (discontinuity), but such disruptions are made more gradually and by smaller steps when controlled by continuity as the first principle. The point is to place a final decision about the mix in the hands of natives who speak in their own interests for those technological innovations which most appropriately serve their local cultures. As Arnold Pacey concludes:

One thing common to most countries where equity has increased alongside economic development is that, in one way or another, the interests of low-income groups have been effectively represented in the political process. . . . Demands have had to be expressed from below before there is a more equitable allocation of resources.[12]

Pacey gives an example of historical continuity.[13] The state of Kerala in south India follows a basic-needs approach rather than rapid-growth. In medical care, for instance, rather than a sophisticated centralized hospital, one of Kerala's notable features has been a well-staffed network of primary health clinics established throughout the state. Largely accessible by walking, the clinics concentrate on child birth, public hygiene, and basic medial care. In the process, operating costs are manageable locally, surgeons and specialized equipment need not be imported from technologically advanced countries. This exemplifies a commitment to continuity, with an element of progress or discontinuity (a drop in infant mortality rates, for instance, to the point of favorable comparison at present with England).

OPERATING PRINCIPLES

With the controlling norm established, one can derive a set of normative operating principles in terms of the major components in historical-cultural formation, normative principles which are technology specific: in politics, justice; for users, openness; with nature, harmony; in industry, stewardship; for education, discovery.[14]

I mean these normative principles to operate simultaneously and not sequentially. While primary responsibility in the educational-scientific community, for example, is discovery, the norms of justice, openness, harmony, and stewardship ought to simultaneously shape the decision-making process. Thus we have expanded our theory to argue that the controlling aim of responsible technology is cultural continuity and its operating principles are justice, openness, harmony, stewardship, and discovery. And, given the overview character of this essay, it is sufficient at this stage to define these terms in the briefest possible manner.

Regarding justice, I refer to distributive justice, to the allocation of resources which respects right relationships in that all participants receive their proper due. Therefore, introducing tools which make work easier while maintaining cultural continuity is objectionable if workers are still exploited and subjugated. In the user domain, open communication is essential if decision-makers are to fulfill their own responsibility regarding economic value, safety, appropriateness, and dependability. Regarding nature, only those technological products which do not violate physical reality and protect non-renewable resources are morally legitimate. Industry, in the fabrication process, ought to operate by the stewardship principle, using human, physical, and financial resources in a non-exploitative manner. On this view, scientists and engineers in particular and educators in general, are expected to invent, explore and create unencumbered by the needs of profit or political power.

In thus establishing a controlling norm and operating principles, we have established a normative theory of two interconnected and irreducible tiers. This theory insists on a straightforward realism which emphasizes rules (what I have called operating principles). Rules guide actions on the psychological and social levels, and outlining them in detail prevents us from operating in a vacuum or toward ends determined by others. However, the theory recognizes the need to ground our rules, and to use law formulas (what I have called the "controlling norm") as the basis for effective technological rules. Given this two-dimensional approach, one seeks in law statements that they be more or less true and in rules that they be more or less effective.

In developing my theoretical model, I have established a master value (continuity) and introduced the relevant operating principles while insisting on their simultaneity. All that remains is reordering this model in terms of that cultural activity we call the technological process. In what follows I demonstrate how in design, fabrication, and then usage, cultural continuity is served and the normative principles operate simultaneously.

TECHNOLOGICAL PROCESS: FIRST PHASE (DESIGN)

I would call the summation of this normative model in the first phase of the technological process, sufficient design. The notion of sufficiency in design is meant to supplant the narrow concept of technical, financial, and marketing efficiency so often used as guides for the development of design specifications. Even from the practical standpoint of product liability, these efficiency criteria, though perhaps necessary, are not adequate for proper specification of a tool or product which must operate in everyday situations. In this light, appliances, for example, would be designed in terms of their dependability, disposability, and reusability, rather than in terms of the manufacturer's economic need for products to wear out so that replacements will be ordered.

A classic case in this regard is the tomato picker designed by the University of California at Davis. The California Rural Assistance League charged that the University violated state law by spending tax dollars to benefit private interests. Nowhere in the design phase was the question asked about the social impact. This classic case is often cited to show how agribusiness was served exclusively, and tasteless, chemically saturated, tough-skinned tomatoes resulted for consumers.[15]

The result of insufficient design is technology moving in an anti-cultural direction. And it complicates matters that the anti-cultural tomato picker is not caused by one person's failure or a single company's greediness for profits or only one profession's moral insensitivity. The university's research and design team did not consider such social consequences as loss of jobs for seasonal workers and driving small farmers out of business. The university administration had no requirements that academic research be conducted in terms of social consequences, since such guidelines were presumed to infringe on academic freedom. And if the university insisted that its agricultural schools must consider the social impact, why should it not make identical policies for its engineering program with military weapons contracts? The legislature supported its state universities with public funds, but in this case the money benefited a few in agribusiness while harming thousands of laborers. Meanwhile, the canneries inked contracts with the huge tomato growers since one reliable supplier seemed more dependable than maintaining relationships with dozens of smaller growers using different methods and scattered geographically. Agricultural research cooperated by developing the right hybrids and adding chemicals for uniform color change which put the growing process on a predictable schedule. Economists calculated that the picker's efficiency would reduce the price of tomatoes, failing to envision that the lessened competition, higher risk, and larger capital investment would dramatically curb incentives to lower prices.

Again, precisely what solutions to recommend is difficult, since so many individuals and units were implicated in this initial phase of the technological process: university engineers, administrators, the legislature, canneries, agricultural researchers, economists. Therefore, the responsibility for sufficient design is not merely a province of engineers. Certainly I do not wish to turn the neutrality view on its head and place all blame on engineers. Following my normative model, however, would result in such changes as smaller machines that do not eliminate all small growers, job retraining for the 35,000 seasonal workers, and adequate testing to prove that the heavy chemicals on the tomatoes and soil were completely safe in both short and long term.

We are repeating the same insufficient design with our work on robotics at present. The social consequences are left out of the engineering phase or discredited as unknowable. Meaningful employment, rather than being a non-negotiable of our domestic policy, is reduced once again to a statistical problem for politicians and does not shape the calculus of engineers.

In summary, my theory of normative technology calls for our replacing the narrow concept of efficiency with a design philosophy of sufficiency in which cultural continuity serves as master value and the relevant normative principles are realized simultaneously.

TECHNOLOGICAL PROCESS: SECOND PHASE (FABRICATION)

The second phase – fabrication – often stands as a testimony of disgrace in industrial countries. As the Frankfurt school recognized, the locus of revolutionary power and class consciousness is no longer the factory, yet Marx's concern for worker alienation and human degradation in the workplace is still of extraordinary importance. No one discounts the progress in wages and benefits since the days when Karl Marx wrote *Das Kapital* in the basement of the London Museum. At least we protect children and have more humane conditions on the whole. Health and safety standards are better.

But with unions on the defensive, high unemployment rates, meaningless work, exploited labor and squalid conditions in our cities, the fabrication phase continues to be outrageous. It is a starkly antinormative in most cases. My complaint centers on a violation of the norm of stewardship, and a failure to recognize the others simultaneously.

Well-worn cases from transportation technology exemplify how product reliability and human safety ought to be non-negotiable, but often are not. The McDonnell-Douglas DC10 was plagued by mechanical problems since it taxied out of its hanger in July 1970 with vice-president Spiro Agnew in the cockpit. With 20 billion in jumbo aircraft sales at stake, McDonnell Douglas was racing to complete its jet before the L1011 and Boeing 747 were finished. But at the cost of being first to gain its market share, the DC10 entered the skies with only 3 hydraulic systems instead of 4, and tires that had not been upgraded from the smaller DC8 and 9 in 1962. Since that inauspicious start cracks in engine mountings appeared, cargo doors blew off, cabin floors buckled, engine fan blades deteriorated in transoceanic flight. Meanwhile – in violation of the stewardship principle – as hundreds of such safety reports were being investigated over a four year period (1974–78), no planes were voluntarily grounded until all defects were corrected.

The B.F. Goodrich brake scandal is another famous story in corporate history, with even the most salutary accounts unable to explain away this company's non-normative behavior. And during the Pinto gas tank episode, Ford executives showed little evidence of wanting a good product, period. They did only what was obligated by law – and in the process disdained the regulators as young imbeciles, do-gooders from MIT.

You need not be a Marxist to bristle at the arrogance, authoritarianism, incompetence of such American manufacturers – as documented, for instance, in Robert Heilbroner's book, *In the Name of Profit*. Johnson and Johnson's response with Tylenol is a refreshing exception, though here the company itself was not perceived at fault and therefore felt less reason to be defensive and obstructionist.

TECHNOLOGICAL PROCESS: THIRD PHASE (USE)

In the user phase, I call for the norm of conviviality – convivial use a shorthand way of saying that the five normative principles are operating simultaneously at this stage. As an illustration, note that the current U.S. policy regarding satellite technology is anti-normative. Satellite technology currently in place is a non-convivial instrument.

While there are more than 4,000 satellites in space at present, only 35 orbital slots are available in fixed, geo-stationary position. The issue is whether orbital positions should be allocated on a first-come first-served basis or whether they ought to be distributed according to the country-by-country principle. The World Administrative Radio Conference (1979) argued that only by reserving spaces for every country can we guarantee equal access to information in the future. The country-by-country principle is also recommended by UNESCO through their MacBride commission that originated from the 1976 Nairobi conference.[16] In contrast, the United States has continued to advocate a first-come first-served policy on the grounds that satellite technology will continue to improve and thereby meet the information needs of the future as more countries demand greater use of satellites.

The debate here is over cultural imperialism. The smaller nations since WW II have sought political independence from colonial powers and economic independence from transnational corporations. In the 1980s they are demanding cultural independence. I mean cultural imperialism in the sense of control over a country's information, the dominance by an industrial country over the films, television and news – the popular culture – of developing countries. This form of colonialism prevents smaller nations from achieving their own identity, from maintaining cultural continuity. When we employ our formal criterion regarding cultural continuity and the normative principles of justice in the political arena, we can only support a master plan which guarantees access country-by-country, reserving positions for future use, rather than the principle of first-come first-served favorable to the United States, Western Europe and Russia.

Injustice here is obvious in that developed countries have almost total technological hegemony over the satellite network – setting advantageous prices, facilities, and service for themselves. Injustice is also seen in the flagrant imbalance of media content. Eighty percent of news received in Third World Countries originates from developed nations. The majority of television shows are produced by developed countries and ill-suited to other nations. In addition, the inaccurate and unrepresentative depiction of non-developed countries – without sufficient means to correct distorted information – is unjust. No international code of ethics, for example, describes a policy for redressing grievances.

Therefore, we must conclude that a new technology such as satellites cannot merely be viewed as technologically neutral, as an efficient machine transmitting global messages faster at less cost. New technologies such as satellites, for all their benefits, do not automatically result in the global unity and greater human understanding suggested by Marshall McLuhan. And clearly Buckminster Fuller's engineering criterion is potentially unacceptable – arguing as he does the virtue of satellite communication because the same volume of information can be transmitted by a technological instrument 1/700 thousandth the weight of the transatlantic cable.

REORDERING OUR LANGUAGE

Hans Jonas' ethics of global solidarity reaches in essence the same conclusion as my theory of normative technology on this question of international communication, but Jonas speaks more compellingly regarding the nuclear threat than of cultural imperialism. Hans Jonas comes close to despair. He refers to the search for a responsible theory of technology as looking "suspiciously . . . like a fool's errand" today. All that remains, according to Jonas, is a "feeling for norm," some intuitive recognition that such things exist.[17]

Given that climate, only by acts of conscience, practical and theoretical, will the smoldering sparks be fanned into flame. In order to operate at all in normative fashion, I agree with the Heideggerian argument for fundamental change in our house of language. Without a radical redress in our symbolic environment, without a vocabulary and vision unencumbered by efficiency, without the overcoming of technicism, normative technology remains an impossibility.[18]

But why turn to the academy – the keepers of language – as the principal agent of transformation? All our institutions stand indicted in some fashion; admittedly, all are complicitors when the technological process turns anti-normative. But I believe those of us in the academy should begin at home, recognizing that normative technology is an empowering idea which can influence other dimensions of the technological process. A critical consciousness sparked by a renaming of the world, as Paulo Freire reminds us, can be the catalyst for a revolutionary praxis.[19]

I am calling for a prophetic witness from those of us interested in the philosophy of technology. I realize the word "prophet" exudes a strong Old Testament aura – Jeremiah, Isaiah, and Elijah calling for repentance from Israel. And in our own day we glibly hail Ralph Nader as a prophet to industry and speak of George Orwell's *1984* as a prophecy of unremitting doom. The prefaces to nearly all of Ellul's newest books call him a prophet to the technological society.

For all those encumbrances on the word, prophet is still the most appropriate term. I refer not to prophets who foretell, who predict the future, but to prophecy as forthtelling. The genuine meaning is speaking a true word, with pathos, with an unquenchable concern for justice – prophets telling the truth strongly and thereby making a permanent difference.

A prophetic witness confronts technicism and insists on desacralizing it. Prophets do not attack technology itself, but technicism as an unacceptable worship of a modern god. Prophets cut through the idolatrous attitudes, intentions, and desires which are driving technology forward. Prophets sever at its roots any blind faith that technological prowess can lead from one achievement to another.

However, while prophetic witness warrants primacy, I do not give it equal ultimacy with the positive aspects of our cultural task. Prophecy must be brought solidly within the circumference of responsibility itself, without making it the epicenter. Evil has no independent life force of its own; it is a parasite living off nature and society. We should make the doing of technology our preoccupation, yet resist the injustices and inhumanities which plague the technological process because of human depravity. Prophetic resistance I regard as a true work, but done with our left hand. I advocate a prophetic witness which encourages a technological process of sufficient design, humanity, conviviality, while witnessing to the darker side of the technological process with tears.[20]

Ivan Illich considers the matter of vernacular discourse central to any significant transformation. He argues that culture is best interpreted as an ongoing attempt by powerful groups to control our common life through monopolizing language.[21] Most Marxist thinkers trace the character of modern life to the rise of big business capitalism, or the development of bureaucracy, or the introduction of powerful technologies. Illich demonstrates that each of these changes rests on a more fundamental historical transformation of the form of communication. The forms of political and economic control since the sixteenth century – be they controls by the Roman Catholic Church, or early capitalism, or twentieth century professionalism – all rest on the domination of language. In each case, Illich argues, powerful groups have expropriated the right to define human needs and how to satisfy them.

Good old words have been made into branding irons that claim wardship for experts over home, shop, store, and the space or ether between them. Language, the most fundamental of commons, is thus polluted by twisted strands of jargon, each under the control of another profession.[22]

Thus doctors come to define for us what health is, lawyers justice, engineers technology, educators intelligence, urban planners cities, and journalists news. By insisting on the symbols of expertise, people are robbed of meaningful participation in these institutions and the social issues surrounding them.

Illich starts with the familiar Marxist argument that both capitalism and professionalism reduce human activities to mere commodities, but he pushes the argument further by suggesting that the two are ultimately linked through the fact that both capitalists and professionals maintain their hegemony by communication. Both create artificial scarcities by symbolically constructing a set of needs that only they can satisfy. Illich thus transforms Marx's vision of history, from one in which a new set of productive relations unfolds in each epoch to one in which successive areas of human consciousness are linguistically colonized by groups in search of wealth and status. Therefore, the urgent need for those who reconstruct our language, and therefore the call for prophetic witness.

Manifestoes, pamphlets, books, educational materials, discussion guides, journals, films – all in the vernacular tongue – are vital to an exercise of prophetic responsibility. Martin Luther's pamphlets, Thomas Paine's tracts, Ivan Illich's books are indisputable evidence that the word is mightier than the sword – that it can cut deeply into our nerve system if operating in the vernacular mode out of a carefully articulated normative worldview.

We ought not to discredit the revolutionary importance of a prophetic witness bounded by the proto-norm of cultural continuity and insisting on the realization of justice, openness, harmony, stewardship, and discovery. Such a prophetic word empowers cultures toward healing and social transformation. And lest the task seem intractable, I am reminded of Dietrich Bonhoeffer's wisdom: "Lay hold of the darkest edges of human frailty, and trim them into decency."

NOTES

[1] R. A. Buchanan, *Technology and Social Progress* (Oxford: Pergamon Press, 1965), p. 163.
[2] Arnold Pacey, *The Culture of Technology* (Cambridge, MA: The MIT Press, 1983), pp. 1–3.
[3] George Grant, "The Computer Does Not Impose on Us the Way it Should Be Used," in A. Rotstein, ed., *Beyond Industrial Growth* (Toronto: University of Toronto Press, 1976), pp. 117–131.
[4] Pacey, *ibid*, ch. 1.
[5] For elaboration, see a similar definition in Stephen Monsma, ed., *Responsible Technology* (Grand Rapids, MI: Eerdmans, 1986), ch. 2. My formulation throughout is influenced by this volume.
[6] Harold A. Innis, *The Bias of Communication* (Toronto: University of Toronto Press, 1951).
[7] Grant, *ibid.*, pp. 125–26.
[8] Pacey, *ibid.*, p. 102.
[9] Quine, *Erkenntnis*, 1975, pp. 327–328.
[10] Hans Jonas, *The Imperative of Responsibility* (Chicago: University of Chicago Press, 1984), p. 6.
[11] Martin Heidegger, *The Question Concerning Technology and Other Essays*, trans. William Lovitt (New York: Harper Torchbooks, 1977), pp. 14 ff.
[12] Pacey, ibid., pp. 76–77.
[13] Pacey, ibid., pp. 70–77.
[14] For elaboration, see Monsma, ibid., pp. 170–177.
[15] Peter Schrag, "Rubber Tomatoes: The Unsavory Partnership of Research and Agribusiness," *Harper's*, June 1978, pp. 24–29; Mark Kramer, "The Ruination of the Tomato," *Atlantic Monthly*, January 1980, 245:1, pp. 72–77; Marjorie Sun, "Weighing the Social Costs of Innovation," *Science*, 30 March 1984, pp. 1368–69.
[16] Sean MacBride, *Many Voices-One World*, (Place de Fontenoy, Paris: UNESCO, 1980).
[17] Hans Jonas, "Technology and Responsibility: Reflections on the New Tasks of Ethics," *Social Research*, Spring 1973, pp. 31–54.
[18] An argument developed brilliantly by Manfred Stanley, *The Technological Conscience* (Chicago: University of Chicago Press [1978], 1981).
[19] E.g., Paulo Freire, *The Politics of Education: Culture, Power and Liberation* (South Hadley, MA: Bergin and Garvey, 1985), chs. 6–8, 12; Paulo Freire, *Literacy: Reading the Word and the World* (South Hadley, MA: Bergin and Garvey, 1987), pp. 1–27.
[20] Cf. Monsma, *ibid.*, ch. 11.
[21] For a cogent review of the argument and literature, see John Pauly, "Ivan Illich and Mass Communication Studies," *Communication Research*, 10:2, April 1983, pp. 259–280.
[22] Ivan Illich, *Shadow Work* (Boston: Marion Boyars, 1981), p. 29.

EDMUND F. BYRNE

GLOBALIZATION AND COMMUNITY:
IN SEARCH OF TRANSNATIONAL JUSTICE

INTRODUCTION

There is a kind of gut-level feeling, especially among those most imme-diately affected, that a plant closing or any other traumatic work-force restructuring must be unjust. But the instincts of corporate decision-makers and their libertarian supporters engender no such feeling. As a result, at least in a capitalist society, anyone who doubts the liceity of this literal form of moving and shaking is assumed to have the burden of proof. So be it, then; the question must be asked: under what condi-tions, if any, is a workforce restructuring unjust?

A deontological answer might be that a workforce restructuring is unjust if the interests of any party involved in the restructuring are seriously undermined. And such an answer might actually prove to be of some polemical use. But its theoretical force is diminished by the fact that these restructurings typically affect a wide range of persons and institutions other than those most immediately and obviously "in-volved," including some who, at least relatively speaking, benefit more than others are harmed. These beneficiaries of the workforce restruc-turing might, in turn, be located far from those who are harmed. Moreover, even if they are members of some society, they may not be in any ordinary sense members of the same society. This is especially the case with regard to workers whose employer is a transnational corpora-tion engaged in globalization.

Globalization is the cumulative consequence of corporate plant loca-tion decisions made on the basis of an assessment of the most favorable business climates in the world, as determined by such factors as govern-ment tax policies and workplace health and safety regulations, infra-structure, the status of unions, the compatibility of available technology and local workplace skills, and local wage levels. In short, it is a global strategy for identifying and exploiting the most owner-advantageous means and relations of production. How, then, against this global background can the justice or injustice of a workforce restructuring be determined?

Edmund F. Byrne and Joseph C. Pitt (eds.), Technological Transformation: Contextual and Conceptual Implications, 141–161.
© *1988 Edmund F. Byrne.*

Among Anglophone social and political philosophers (at least those who think of themselves as liberal) it is widely assumed that questions of justice are best handled by applying some version of a social contract theory, preferably that of John Rawls. And at first glance one might assume that Rawls' conception of justice as fairness is surely applicable; but Rawls discourages a global agenda by blurring (not merely neutralizing) the boundaries of a society with regard to both time and space.

With regard to *time*, Rawls' contractors are expected to arrive at principles of justice without knowing to which generation they belong or to what level of civilization and culture their society has attained. But the history of conquest and enslavement is enough to indicate that the variable of time is relevant to justice with regard to workforce restructuring. With regard to space, Rawls simply fails to consider the bearing upon a theory of justice of a society's boundaries and its degree of penetration into or by another society or societies (leaving all of this unquestioningly to international law). But technology transfer in general and plant relocation in particular frequently involve a process the justice of which can be determined only if its impact on two or more geopolitically distinct *places* is taken into account. In order to establish conditions of justice with regard to workforce restructuring, then, one needs to take into account certain spatiotemporal facts, i.e., facts about the geopolitical position in place and time of those involved.

In this regard, a number of philosophers have called attention to the moral irony of espousing universal principles of conduct but confining their practical import within a nation-state.[1] But few have found a consistent or convincingly less noxious way to make the globe the geographical correlate of universality. Especially obstructive for defenders of a global ethic is, as social contractarians regularly note, the manifest inadequacy of global enforcement of standards. This, however, is an instance of what Brian Barry calls the compliance problem – important in its own right but subsidiary to the problem of establishing universal, read: global, norms.[2] A perhaps even more serious objection is that a global interpretation of justice tends to disregard the moral significance of loyalty to less than global groups, be they families or nation-states or whatever. This disregard is especially suspect (although on occasion justifiable) when it involves intrusive imposition of some exogenous standard of behavior.

The most promising solution to these problems will be found, I believe, in Haskell Fain's carefully wrought task-oriented approach to problems that require cooperation across national borders.[3] But his groundbreaking enterprise deserves full attention in its own right. For the project at hand, namely, finding a way to articulate moral scope, we can get along with Julius Stone's concept of a justice-constituency, which involves "the claimants and beneficiaries of justice in concrete times and places, with their biological endowments and social environments, and their limited and tentative envisionings of the future." A justice-constituency, he says, is typically coterminous with a nation-state but might just as well be a "postulated justice-constituency of mankind."[4] I submit, in light of problems regarding work, that it might also involve a social unit less complex than a nation state.

Concerns similar to those about globalizing universal principles obtain also in the case of workforce restructuring. For, to the extent that only global strategy is deemed to be at issue, and the denationalized leaders of a nation-state are compliant, the interests of more localized justice-constituencies will be disregarded. With this in mind, I shall use the term 'community' to mean abstractly just such a geographically localized complex of legitimate interests and concretely the human beings who assign these interests moral priority.

It would, of course, be an oversimplification to say that each and every workforce restructuring involves two communities, e.g., one which experiences a plant closing and the other a plant opening. But this very oversimplification does suggest a device for describing the often conflicting interests at stake in a transnational restructuring, namely, as a kind of voyage. Imagine, if you will, that those harmed are, as it were, at the "point of departure" of a plant relocation and those somehow benefiting are at the "point of arrival." With this imagery in mind, I shall consider first communities at a point of departure of a workplace restructuring (exemplified by situations in the United States) and then those of communities at a point of arrival. My objective is to suggest (1) that community interests constitute a moral and arguably also a legal basis for limiting globalization; and (2) that justice as fairness can be meaningful in the transnational arena only if it requires a community-oriented limit on corporations.

I. COMMUNITIES VERSUS CORPORATIONS: THE AMERICAN EXAMPLE

Even the staunchest defenders of American business generally concede that some workforce restructuring over the last several decades may have harmed some communities. They excuse this harm, however, by appealing to "business necessity." Against this excuse, communities have little recourse, especially when the decision in question is said to be required to meet global competition. Even unions, however reluctantly, tend to be cooperative. Unemployment and social disruption are inevitable side effects. But whether plants are closed or automated, that business necessity is an adequate justification goes largely unchallenged.

Take plant closings, first of all. Many labor-intensive plants have been closed in recent years in the United States, especially in areas that developed industrially at an earlier stage in our history: the so-called "rust belt." Why is this the case? Some blame rising labor costs (hence the emphasis on wage "concessions" in the 1980's). Others, including experts at the International Labor Organization (ILO), prefer to blame "the importance of technological innovation as a means of (meeting) competition".[5] The pressure of competition may motivate a desire to innovate. But it may just as well inspire companies to favor environments where cheaper labor is available even over those where innovation might most easily be introduced. The consideration that tips the balance may be an opportunity to "get out from under" a union.

Bumper stickers say "Buy American" and (at least until the decline of the dollar) unions had hopes of getting some sort of "domestic content" legislation passed. But what is there "American" to buy? Most of the components of an IBM-PC are manufactured abroad. Ford Motor Company prides itself on its "world car," the multinational manufacture of which is intended to be invulnerable to any local labor unrest. Such practices are duplicated by hundreds of companies. In autos, for example, the U.S. Department of Commerce estimates that by 1988 17% of the cars sold by Detroit will be produced abroad, at the cost of an additional 90,000 lost jobs; and by 1995 29% of the parts in domestically produced cars will be imported, with comparable worker displacement as a consequence. As a result, one domestic auto plant after another is scheduled for closing, and communities ponder the legality of suing for the institutional equivalent of alimony.

This kind of story can be repeated with regard to industry after industry. Where once we had manufacturing strength we now might have what *Business Week* has called "the hollow corporation": a service-oriented company that does everything from designing to distribution, but whose products are manufactured abroad.[6]

What will be the long-term results of this transformation of our economy? Some observers see the shift from manufacturing to service as a progressive move beyond drudgery. Others, including leaders of Japanese industry, watch in disbelief as we lose not only money but, arguably, the longterm ability to make money.[7] Such concerns, however, are salient only if one is prepared to take into account interests other than those of financial investors when formulating industrial policy. To a stockholder it matters little where a company's profits have been generated. Such geographical details are important only to those whose livelihood is locally constrained: the would-be employees and host community of a profit-making enterprise. As even middle management personnel are coming to realize, the rights of these constituencies are limited at best. But why should stockholders be the only investors who matter?[8] Employees also make investments in a worksite, as does the community. Should not such investments be taken into account when a company contemplates the future of a worksite?

As it is, businesses come and go, according to the latest paradigm of a "good business climate." Typically, they concede neither to workers nor to surrounding townspeople any effective control over the terms and conditions of their stay. Whatever the issue, whether it is workplace health and safety, wages and benefits, imminent layoffs or even a plant closing – secrecy is sacred in corporate America; and so long as other corporations are not thereby shut out of the market, courts consider this to be an entirely appropriate way of doing business. Absent some "right-to-know" legislation, they seldom agree that workers might actually be entitled to "management level" information that affects their jobs.[9]

Consider, secondly, the impact of automation. For several decades, the availability of much cheaper labor abroad has somewhat blunted management's interest in automation as the key to making profits without payrolls. But some analysts of the corporate exodus to foreign shores see problems ahead; so they are looking with renewed interest to automation as an answer to the pitfalls of globalization. As this view was recently expressed in *Business Week*:

(T)he new technology would bring a crashing halt to the madcap chase after cheap foreign labor. Because labor costs would be virtually zero and other offshore savings – cheaper materials and lower overhead – would be overwhelmed by the benefits of quick turnarounds and low inventories, the computerized factory could produce things in Indiana for less than they cost to import from India.[10]

Disregarded in this projected solution is any concern about workers. In fact, this very silence on the subject of workers exemplifies the traditional attitude of management that labor is simply a cost of doing business and so should be reduced as much as possible. When skilled workers in the nineteenth century challenged the impoverishing terms offered to them by factory owners, the owners responded by seeking a technological substitute for their unruly employees: what Scottish engineer Andrew Ure called "the automatic plan." Today, however, the target is no longer only the comparatively insignificant direct, or "touch," labor costs (10–15% in the U.S.) but additional factors that account for three times as much of the cost of production: indirect labor, middle management, and other overhead.[11]

In the abstract, the purpose of workforce restructuring is to save labor, i.e., to save a company some of the costs associated with paying labor. Concretely, workforce restructuring is inherently, even if not intentionally, hostile to anyone whose labor is thereby "saved." So it is easy to see how an a posteriori analysis of this process can lead to the claim that management tilts cost-cutting decisions whenever possible in the direction of cutting payrolls.[12] If this is meant as a criticism, however, it usually falls on deaf ears in corporate headquarters, legislatures, and courtrooms. Under American labor law corporate cost-cutting is just good business, unless (in some situations) it is found to have been motivated by anti-union animus. Not surprisingly, then, secondary effects of a restructuring on the host community are treated as a subject for regret but not for liability.

The corporation/community issue here at stake is easy to illustrate in a rust-belt state like Indiana, which has lost some 18 percent of its manufacturing jobs in the 1980s. Consider just two plant closings, each by a "model" company.

The first involves IBM, which for years was a major employer in the quiet college town of Greencastle. Recently, as a part of corporate restructuring in response to global competition, it closed its Greencastle facility, a parts warehouse. To soften the blow, it offered transfers to its employees, many of whom were long settled in the community; and in an act of corporate generosity it offered to "donate" the warehouse facility to the city and "lend" it one of its managers to assist in picking up the pieces. In the year following the closing, the city in fact persuaded a number of smaller, and more diversified, companies to locate there. The results are encouraging in comparison to the debacle initially contemplated. But new employees are not generally as highly skilled or compensated as their predecessors, the value of vacated real estate is at risk, and no one is talking about what it cost the community to attract IBM in the first place.[13]

Columbus, Indiana, is the world headquarters of Cummins Engine Company, long renowned as a model of the socially responsible enterprise. While providing more than half of all diesel engines sold for trucks and heavy equipment in North America (but only 2–3% of the global market), Cummins paid a world renowned architect to design each of thirty major buildings in Columbus. Of its 19,500 employees worldwide, as many as 11,000 were on the payroll in the Columbus area as recently as 1980 (including, incidentally, a business ethicist). Then came the worry that a Japanese competitor would be moving into its bailiwick. Management responded by launching a $1 billion program to expand its product line, just when a worldwide recession dropped its engine sales from 125,000 in 1981 to only 85,000 in 1982. The next three years it sought, and achieved half of, a 30% reduction in costs via greater efficiency from suppliers and employees. In the third quarter of 1986 the company reported a loss of over $100 million, at which point management announced the closing of two Columbus area facilities to effect a 45% reduction in its local workforce. Most of this has been accomplished by firings and forced early retirements – not of unionized hourly workers, but of union-"exempt" middle management employees. Those fired who agreed to sign a waiver not to sue were given several months' severance pay. A dismissed employee who did sue was stopped short by a summary judgment in favor of Cummins, employment-at-will still being the unqualified rule in Indiana. Those still employed along with the rest of the community continue to find positive things to say about Cummins, but the myth about its singularly benevolent role in their lives is gone forever. Company-community relations will never be the same again.[14]

Such top management responses to "business necessity" are not aimed at mere profitability. Plants so targeted may well be profitable without being moved or automated. But the investor-oriented bias of the corporation requires management to seek not just profit but maximum profit. (Cummins Engine, for example, has not lost any of its market share; and now that the dollar has weakened somewhat relative to the yen, the company's recent workforce reduction seems much less inevitable, unless perhaps the company intends to relocate the closed facilities in a cheap labor setting abroad.) It is in its quest for maximum profit that even the most benign corporations in America routinely interpret the right of private property as a license to use or abuse a community as it sees fit.

License to use or abuse a community extends the old Roman rule that the owner of property has the right to use or abuse (*jus utendi et abutendi*) *just that property* as the owner sees fit. On the other hand, since the rise of democratic states an owner's control over property has in fact been limited by various social constraints. Imposed at least in principle for the public benefit, these constraints can be detrimental to the interests of an individual property owner. But they seldom interfere with a major corporation, whose use of property is limited only by "business necessity." This can be seen, for example, in a 1987 decision of the U.S. Supreme Court[15] which can be interpreted as saying that government must compensate a private owner for even minimal interference with the owner's use of property. But, as we shall see, other decisions even from the same court have been more attentive to community interests.

The interests of a community in a corporation can, of course, be understood to include various components. Foremost among these, according to some financiers, are the stockholders, as was emphasized by another 1987 U.S. Supreme Court decision that authorizes a state to legislate a cooling off period before raiders can acquire control of a company.[16] Also to be taken into account, however, is the investment of the workers, as the NLRB argued in *Ozark Trailers, Inc.*,[17] and as is becoming more apparent in the wake of legislation that facilitates worker ownership of companies (ESOP's). Similarly, local taxpayers are investors through whatever "industrial development" arrangements have been made in their name. The interests of these components of the community should not be overlooked as local governmental units concern themselves with accommodating a major corporation.

Attention to community interests may be discerned from emerging law that imposes limits on the owners of private property, (1) by restricting their use of that property and (2) by public preemption of private property. The first involves what is called inverse condemnation; the second, eminent domain. Each involves what is technically known as a taking, because of the relevant language in the Fifth Amendment to the U.S. Constitution: "(N)or shall private property be taken for public use, without just compensation." (Satisfactory resolution of the difficult and controversial question of just compensation will here be assumed.[18])

The Fifth Amendment of the U.S. Constitution requires that the federal government and, via the Fourteenth Amendment, state governments (or subsidiaries thereof) take private property only for "public use." As applied particularly to eminent domain, there was concern from earliest times that public use might be interpreted to mean nothing more than public benefit. There was good reason for such concern; but in retrospect even that standard would be immeasurably superior to what has been the de facto standard, namely, benefit to corporations.

This abdication of public responsibility commonly takes the form of appellate deference to the decisions of local courts. Such deference would be constructive if local courts were fora in which a community's true interests are deliberated. But these courts are usually under pressure to approve the pet project of one or more corporations cloaked in the mantle of "public use." The programmatic significance of the following analysis, therefore, depends on the extent to which a truly participatory community is able to have its interests impartially adjudicated. For, the letter of the law is fairly straightforward; what is crucial is how "public use" is interpreted.

A concurring opinion in an 1837 New York State case[19] worried about the implications of identifying public use with public benefit because this would diminish the property rights of the individual against whomever the government represents. This equation, however, prevailed both in the East[20] and in the West.[21] But in the Midwest, courts tended to favor a much more conservative approach to taking, limiting public use to what is absolutely necessary.[22] (This, as we shall see, has changed.)

In 1936 New York courts found that urban renewal constitutes a public use even if some structures affected are not substandard and some private firms are involved in the project: *New York City Housing Authority v. Muller.*[23] The *Muller* reasoning became a national standard in 1954 when the U.S. Supreme Court endorsed it in *Berman v. Parker.*[24] Since then public benefit has, to a considerable extent, become virtually indistinguishable from private, i.e., dominant corporate, benefit. For example, a New York court saw nothing improper in the condemnation of thirteen city blocks and all the businesses, however thriving, on these blocks: *Courtesy Sandwich Shop, Inc. v. Port of New York Authority* (1963);[25] and a Michigan court blithely upheld the condemnation of an even more extensive urban neighborhood so that General Motors could allegedly provide jobs and taxes by constructing a new plant: *Poletown Neighborhood Council v. City of Detroit* (1981).[26]

Happily, not all eminent domain cases follow the path of least resistance to corporate fiat. For example, in *Hawaii Housing Authority v. Midkiff* (1984),[27] the United States Supreme Court upheld Hawaii's use of eminent domain to redistribute land that had been monopolized by a small minority of owners, resulting in artificially high prices on what little land was available for purchase.

Deferring as usual to the state government's prerogatives in these matters, the Supreme Court found the Hawaii legislation to be a "comprehensive and rational approach to correcting market failure."[28] This was enough, because

where the exercise of the eminent domain power is rationally related to a conceivable public purpose, the Court has never held a compensated taking to be proscribed by the Public Use Clause.[29]

Decisions such as these are encouraging to one who favors recognition of the interests of the community in law. But they have not dislodged a traditional unwillingness to consider property other than real estate in the same way. This was shown, for example, in the Court's recent denial of certiorari to the "plant closing" case of the (now) Los Angeles Raiders, who moved their franchise but not their stadium from Oakland.[30] Some legal scholars, such as Charles Reich, Frank Michelman, and Joseph Sax, have argued that even entitlements, e.g., under Social Security, should be protected as property.[31] (This strongly communitarian step the Supreme Court has been unwilling to take.[32]) In the same vein, the law of public use could provide a basis for a more communitarian assessment of plant closings and new technology. But the precedents are ambiguous at best, especially where corporate priorities are at stake.

II. COMMUNITY VERSUS GLOBALIZATION: THE "OFFSHORE" CASE

If accused of unjustly restructuring a local workforce, a transnational corporation might defend its decision by appealing to a global concept of justice. In the abstract this appeal is at least prima facie persuasive. On utilitarian grounds it might even be beyond challenge if the calculus of benefits shows that harm done to a community at a plant relocation's point of departure represents a comparatively insignificant component of otherwise advantageous consequences. But a moral account might be less supportive if it considered more carefully the effects on communities at the point of arrival.

Having argued already that the interests of a community at the point of departure should constitute a limit on globalization, I shall now argue that globalization should also be limited by the interests of a community at the point of arrival. For this purpose one might want to move Rawls' principles of justice out of their unisocietal confines onto the complex matrix of the multisocietal globe.

Rawls' Kantian interpretation of the original contract is unquestionably a monumental accomplishment, and will surely continue to inspire learned progeny for generations to come. But it fails to establish what constitutes a society and assumes (by its acquiescence in the conventions of international law) that a society may be just in relative isolation from the rest of the world. Having so limited the scope of his treatise, Rawls erects two obstacles to the development of a contractarian theory of global justice. First, his ahistorical presentation avoids rather than resolves Hume's contention about the origins of nations-states. "(A)lmost all governments . . . ," he said,

have been founded originally, either on usurpation or conquest, or both, without any pretence of a fair consent or voluntary subjection of the people.[33]

Secondly, he perpetuates Rousseau's indifference to corporate structures and the impact that can have on an agrarian symbiosis between a population of human beings and the land they inhabit. According to Rousseau,

The ideal ratio (between territory and population) is achieved when the land can support its population, and when the population is of a size to absorb all the products of the land.[34]

Any human ecosystem thus circumscribed is, as Rousseau well realized, seriously threatened by incursions from without; hence, he added, that people is best suited for laws "which is not dependent upon other nations, nor needed by them."[35]

Such isolated autonomy exists only in the abstract. In the actual world, inter- and trans-societal interaction is ever possible and, in a technologically advanced global village, inevitable. What matters, as far as a theory of justice is concerned, is how the terms and conditions of such interactions are established. I take it as given that justice cannot be assured if not all the interested parties are adequately represented in making this determination. Still less can it be assured by a hypothetical contract the parties to which are oblivious of their own community's interests. This can be seen by imagining Rawlsian contractors who are asked to arrive at (a) principles of justice and (b) a constitution without knowing the specific space-time coordinates of the society with which they are in fact associated. These preliminaries out of the way, let the social contracting commence.

The contractors meet, not knowing their generation or their place in society; but they do know that they belong to a particular society (a remarkable grant of wisdom to which many people in the actual world attain only with great difficulty if at all). Thus minimally informed about relevant details, they ascend beyond utilitarianism to transcendental rationality and arrive at Rawls' principles of justice as fairness. The veil of ignorance is then partially lifted for a constitutional convention. "They now know the relevant general facts about their society, that is, its natural circumstances and resources, its level of economic advance and political culture, and so on."[36] With only this additional knowledge, Rawls assures us, they manage to draft a constitution; whereupon they reenter the society they have wisely ordered.

So much for Rawls. Now for the rest of the story – or, rather, stories, since the variables cover an open range of possibilities, most of which are fraught with ambiguities. Consider for purposes of illustration just a few actual space-time situations to which the wisely devised ordering principles are expected to apply. In none of them, it may be noticed, is there anything approaching consensus among the inhabitants about what constitutes their society, what everyone's place in it is or, what is still more basic, who is or is not included in the society.

1. Palenque, province of Chiapas, in southern Mexico, 1987. You are a Mayan. Centuries ago your ancestors were dominant in this part of the world, and relied upon human sacrifice, among other things, to maintain their preeminence in the world. But long ago they were conquered by Zapotecs, then Toltecs, and then definitively by Spanish conquistadores. Impressive ruins outside of town suggest the former greatness of your people. But you yourself live in extreme poverty. You and your people are at the margins of the prevalent society in Mexico, since its values are based on the imported European culture, which attributes no value to anything "Indian" unless it is respectably pre-Columbian. You do not have reliable running water, so you risk dying of some form of intestinal disease. Besides, advancement of the lumber and oil industries in decimating the rain forest, and thereby destroying the water table itself. Ironically, the resulting desiccation is also destroying the ruins of your ancestral city.[37]

2. Cape Breton Island, Nova Scotia, 1976. You live in the poorest part of Canada, the Maritime Provinces, whose people have for generations felt, with good reason, that they are neglected by the federal government. Like others before you, you and your family have (until now) managed to eke out a living by subsistence farming and fishing. Now you are on the verge of losing your land because you cannot afford the rapidly increasing property tax. The increased value of your land is due to the construction nearby, in the Strait of Canso, of a deep port to accommodate tankers that deliver petroleum from the Middle East to be refined here for direct sale and for use in petrochemicals. The facility itself was built almost entirely by imported labor using materials prefabricated in the United Kingdom.

3. East Timor, 1976. You are one of the more advantaged mountain people who are being systematically slaughtered by Indonesian troops, armed and supplied by the United States, that have invaded this country (once under Portuguese control) to preserve it as an enclave for Western democracy. Your life expectancy is minimal because someone has told the soldiers that a member of your family favored the opposition party. Your interest in this territory is deemed expendable in view of its strategic importance to the oil industry, both because tankers pass nearby and because there are believed to be significant deposits of oil offshore.[38]

4. Republic of Korea, (South Korea), 1980. You are a thirteen-year-old female. In dutiful obedience to your father, who exercises patriarchal authority over all members of his family, you are working in a foreign TNC-controlled electronics assembly plant in the Masaan Export Processing Zone (EPZ). Ninety percent of your fellow workers are young females like yourself. You work eighteen hours a day, seven days a week, for less that $18/month. Throughout your working day you peer through a microscope as you work to meet your quota. Although you wear gloves, you routinely suffer acid burns on your fingers. Within a year of your initial employment you had suffered permanent vision damage. Soon you will no longer be able to do this work, which has provided you with no skills that are transferable to other kinds of work in your country.[39]

5. Taipei, Taiwan, 1987. You used to work on an assembly line in a converted upstairs apartment with dim light and no air conditioning; but now you are a bank employee. Your husband is a quality-control manager at a textile company. Together you earn $1,500 a month and, after paying all expenses, have $500 left. But your extra money is a problem for you, because consumer goods are expensive, credit is not readily available, ordinary interest is very low and the domestic stock market is highly speculative. Your country's economy is already almost totally dependent on foreign-based, especially American, transnational companies. These companies typically exploit cheap labor in Taiwan to manufacture products for export back to their domestic market. As a result, America's trade deficit with Taiwan ($15.7 billion in 1986) is due largely to products made in Taiwan for such companies as Sears, K-Mart, J.C. Penney, Wilson Sporting Goods, Mattel, Schwinn, Hewlett-Packard, IBM, Texas Instruments, Digital Equipment, and – Taiwan's biggest exporter – General Electric.

The American government is pressuring Taiwan to be more accommodating to American imports, but each Taiwanese would have to spend an average of $7,500 (twice the average per capita income) on American goods to overcome the trade deficit. Partly because of the protectionist policies of the Taiwanese government, your country's cash reserve ($35 billion in early 1987) will soon be the largest in the world. If the Taiwan currency is not significantly devalued, foreign companies may begin placing their orders elsewhere. Meanwhile, the political autonomy of your country vis-à-vis mainland China remains unresolved.

Under the circumstances, you and your husband agree with others of your generation that you might as well spend your money while it is still worth something.[40]

6. The West Bank of the Jordan River, 1987. You are a Sephardic Jew. You came to Israel from a nearby country, where you were oppressed, with the understanding that Israel is your "promised land." You have gone to live on the West Bank. This territory is claimed both by Israelis and by Arabs. You have no skills that are marketable in this economy and have been unemployed since your arrival here. Meanwhile, more than 100,000 Arabs come into Israel every day to perform menial jobs. Being a Sephardic Jew, you are of marginal interest to the European Jews who control Israel and to your Arab neighbors. Within another generation the majority of the population of Israel will be Arabs. But only Israeli Jews are granted citizenship. Meanwhile, comparatively few Israelis are yet willing to grant separate political status to this territory, just as Arab nations refuse to acknowledge the existence of Israel as a state. This volatile mixture called Israel is supported by $3 billion/year in aid from the United States.[41]

7. *Dominican Republic, 1980.* You are an employee of a United States-based transnational corporation which operates a plant in the La Romana Economic Free Zone (EFZ), which is essentially a regime of exception from regulation by the national government. You are paid 50 cents an hour, with capital raised locally in support of the enterprise. Soldiers and barbed wire fences around the plant prevent potential union organizers from entering.[42]

8. *Tupecamaru, outside of Lima, Peru, 1987.* You are a direct descendant of the once powerful Incas, about whose language and customs you know little (unlike your relatives in the high country). You are a poor entrepreneur. Under existing law you cannot incorporate, obtain credit or title to the land on which you operate your business. Your home was built illegally on land to which you cannot get legal title. The "informal sector" of the economy, of which your endeavors are a part, accounts for 60% of the total economy and owes nothing, whereas the "formal sector" accounts for only 40% and owes $11 billion. In fact, illegals are responsible for a total of $8.7 billion worth of housing, including more than half of the homes built in Lima. Not being recognized by the government as a community of human beings, however, the illegals are not provided with water or electricity. But they have themselves erected a water tank and now the government has run an electric line to operate a pump – thus, arguably, officially recognizing the existence of the illegals. But half a million laws and regulations, devised mostly by unelected bureaucrats, continue to protect the vested interests of the country's old families.[43]

9. *Northern Mexico, 1987.* You are in any one of eight cities bordering on the United States – Tijuana, Mexicali, Nogales, Piedras Negras,

Ciudad Juarez, Nuevo Laredo, Reynosa or Matamoros. Officially, your town is part of the United States of Mexico. But your town is for all practical purposes controlled by the *maquiladoras*, i.e., border factories built by American-based companies to process raw materials for exportation into the United States as finished products. The factories have come here to take advantage of cheap labor. If you are a very young Mexican female, you may be employed in one of the "in bond" factories (since 90% of their employees are so characterized). If you are a Mexican male, you are probably unemployed; but if you are married you may console yourself with your traditional patriarchal perquisites. You may also be involved in a secondary occupation, such as prostitution or smuggling workers, drugs, oil or cattle to the north or consumer goods or military equipment to the south, e.g., to bolster some revolutionary movement. If you are in any of these groups, you were at best poorly represented behind the veil of ignorance, since you have little freedom and less equality of opportunity. In fact, considering the likely consequences, you are probably worse off if you are employed in a factory than if you were not.

On the other hand, you may be a North American plant manager, in which case your life style is very atypical. You do not speak Spanish, so you require your employees to learn English, in U.S. border towns. You are obviously benefiting from these arrangements, as is your company and its investors. But it is difficult to identify a society in whose founding social contract you might have been represented. Still less would Japanese companies have been represented, yet they are apparently on the verge of imitating and perhaps surpassing your company's exploitation of this advantageous labor market.[44]

10. Los Angeles, California, 1987. In the interest of closing the circle (and thereby modifying the rigidity of the oneway-trip model), imagine finally that you are an undocumented worker – from Mexico or, possibly, the Philippines – employed as a seamstress in a single-owner textile "sweatshop." In return for a 10-hour day of sewing in minimally tolerable surroundings you are paid, maybe, $2–3.00. To supplement your meager pay you might take home a bag of additional cloth on which you work well into the night – because you need the money. Obviously you would like to earn more, but you really have no other viable option. This piece-work industry is able to compete with automated textile plants only by keeping wages to a minimum. Besides, you are in a double bind with respect to your residence in the United States. In your home country you and your dependents were impoverished. In the United States, under recently enacted legislation, the Immigration and Naturalization Service is authorized to fine your employer and deport you just for being where you are; secondly, under a law passed to preclude exploitation, you are subject to penalty for sewing in your home; and, thirdly, you are subject to the animosity of unionized textile garment workers who resent your willingness to work for less than minimum wage. So it is as much in your interest as it is in that of your employer that you call as little attention to yourself as possible. Needless to say, the choices you have made had little to do with your or anybody else's sense of what is just.

What these actual space-time situations illustrate is that the traditional concept of property rights allows for the most unbridled injustice in the world if it is not somehow circumscribed. This circumscription might utilize in a global context Rawls' requirement that a societal system must be so arranged that no other would be more advantageous to the least advantaged group. For, on a global scale, the least advantaged group is made up of disenfranchised (potential) workers. But the results of such an approach would need to be compared with the more demanding collectivization of property that Marxist global justice would presumably require.

In both its domestic and its foreign manifestations, then, corporate enterprise still remains essentially free of the kinds of social responsibility that some jurists, philosophers and political scientists espouse.[45] The ontological rationale for this corporate autonomy requires at least the assumption that a corporation exists in a dimension of the universe that is totally isolated from any community where it happens to have located a facility. Although widely accepted without question, this assumption is manifestly self-serving. It is undercut, however, by each of four conflicting claims regarding ownership and control of property.

The earliest conflict was that between management and labor. In our own times, some recognition is being given to worker ownership, e.g., by means of ESOP's in the United States . But as is often noted, mere ownership does not necessarily include control.

The second conflict is that between stockholders and management. Received doctrine has it, of course, that the stockholders are the true owners of the companies whose stock they own. But until recently management has exercised effective control over most companies, at least those that are publicly held, except, for example, when a takeover bid proves successful. The growing influence of institutional, including foreign, investors may eventually diminish the power of management.[46]

The third conflict is that between a corporation and a host government. Here again, as leaders in developing countries have learned, merely owning a controlling interest in a business is of little value if one does not also control the political and economic environment, often global, in which that business is conducted.[47] But this problem is not in principle any more insoluble than was, say, that of emancipation when slavery was widely presumed to be an appropriate application of the received doctrine regarding ownership of property.[48]

The fourth conflict, on which I have focused here, is that between company and community. As one long trusted company after another resorts to "downsizing" and corporate flight, the loyalty of their employees, even those on the management level, is being severely undermined.[49] Equally disillusioned are communities that had come to think of these enterprises as permanent fixtures in their midst. But why should communities have been so trusting in the first place? For, it is often they that have been controlled by the corporations, and not the other way around. The Pullman Company, for example, literally owned the town of Pullman, Illinois, and ruled it like a feudal lord; and, in a somewhat more subtle way, the rubber companies controlled Akron, Ohio, just as the steel mills controlled Gary, Indiana; and transnational corporations, like colonizers of old, often control the communities where they locate a plant.[50]

Such unilateral corporation/community relationships are common; the relationship characterized by truly cooperative attention to mutual interests is still rare. But social justice surely requires no less. In particular, the right of a corporation to operate globally should be strictly conditioned on its consistently demonstrated recognition of the interests of communities no less than those of stockholders; and this is especially the case when the community in question is, or is on the verge of becoming, a representative of the globally least advantaged.

NOTES

[1] See Anthony Ellis, ed., *Ethics and International Relations*, Manchester: Manchester Univ. Press, 1986, especially James Fishkin, "Theories of Justice and International

Relations," pp. 1–23; Brian Baxter, "The Self, Morality, and the Nation-State," pp. 113–26. See also Charles Beitz, "Cosmopolitan Ideals and National Sentiment," *J. of Philosophy* 80 (1983) 591–600; and Edmund F. Byrne, "The Depersonalization of Violence," *J. of Value Inquiry* 7 (Fall 1973) 161–72.

[2] "Can States Be Moral? International Morality and the Compliance Problem," in Ellis, *op. cit.*, pp. 61–84.

[3] See Haskell Fain, *Normative Politics and the Community of Nations*, Philadelphia: Temple Univ. Press, 1987.

[4] "Approaches to the Notion of International Justice," in Richard A. Falk and Cyril E. Black, eds., *The Future of the International Legal Order*, Vol. I, *Trends and Patterns*, Princeton, NJ: Princeton Univ. Press, 1969. See also Thomas Scanlon, "Contractualism and Utilitarianism," in *Utilitarianism and Beyond*, eds. Amartya Sen and Bernard Williams, Cambridge: Univ. Press, 1982, pp. 103–28.

[5] Gus Edgren, "Employment Adjustment to Trade under Conditions of Stagnating Growth," in D.H. Freedman, ed., *Employment Outlook and Insights*, Geneva: International Labour Office, 1979.

[6] *Business Week*, March 3, 1986, p. 58. Regarding semiconductors in particular, see "Is It Too Late to Save the U.S. Semiconductor Industry? *Business Week*, Aug. 18, 1986, pp. 62–7; Thomas G. Donlan, "Can This Be Silicon Valley?" *Barron's*, March 30, 1987, p. 8 ff..

[7] See, for example, Stephen S. Cohen and John Zysman, *Manufacturing Matters: The Myth of the Post-Industrial Economy*, New York: Basic Books, 1987.

[8] See Adolf A. Berle and Gardiner C. Means, *The Modern Corporation and Private Property*, rev. ed., New York: Harcourt, Brace & World, 1968 (first published New York: Macmillan, 1932).

[9] A pertinent illustration of this attitude is the paternalistic rationale for secrecy in *General Motors Corp., (GMC Truck & Coach Division)*, 191 N.L.R.B. 951, 952 (1971). Included in the Omnibus Trade and Competitiveness Act of 1987, currently pending before the U.S. Congress, is a requirement that employers give employees 60-day notice of any planned closing or layoffs.

[10] *Business Week*, March 3, 1986, p. 72.

[11] *Business Week*, June 16, 1986, p. 101. Ure's expression will be found in his *The Philosophy of Manufactures*, London: C. Knight, 1835, p. 20.

[12] See Edmund F. Byrne, "The Laborsaving Device: Evidence of Responsibility?" in *Philosophy and Technology* IV, ed. Paul T. Durbin, Dordrecht/Boston: Reidel 1988, pp. 63–85.

[13] *Indianapolis Star*, *passim*, July–Dec., 1986.

[14] *Indianapolis Star*, July 5–8, 1987.

[15] First English Evangelical Lutheran Church of Glendale (Calif.) v. County of Los Angeles, U.S. Supreme Court, 85–1199, June 1987. See "Court Tilts Scales towards Property Owners, *National Law J.* 6/22/87, p. 5 ff. See also Noel Peirce, "Placing Constraints on Property Rights," *Indianapolis Star*, 6/21/87, p. F7.

[16] CTS Corp. v. Dynamics Corp. of America, U.S. Supreme Court, 86–71, April 1987. See "Takeover Artists Take a Direct Hit," *Business Week* 5/4/87, p. 35; Stephen Labaton, "For the States, a Starring Role in the Takeover Game," *NY Times* 5/3/87, p. F8; Kirk Victor, "States Flex Muscles on Takeovers," *National Law J.*, 6/1/87, p. 1 ff. See also

Paula Dwyer, "Merger Mania: the courts finally start looking out for shareholders," *Business Week* 11/11/85, p. 35.

[17] 161 N.L.R.B. 561, 566 (1966).

[18] See William Michael Treanor, "The Origins and Original Significance of the Just Compensation Clause of the Fifth Amendment," 94 *Yale L.J.* 694–716 (Jan. 1985).

[19] Bloodgood v. Mohawk Hudson RR Co. (N.Y. 1837), 18 Wend. 9.

[20] Scudder v. Trenton Delaware Falls Co., 1 N.J. Eq. 694 (1832).

[21] See, e.g., Fallbrook Irrigation District v. Bradley, 164 U.S. 112 (1896).

[22] See. e.g., Ryerson v. Brown, 35 Mich. 332 (1877).

[23] 270 NY 333, 1 N.E.2d 153.

[24] 348 U.S. 28.

[25] 12 NY2d 379, 190 NE2d 402.

[26] 410 Mich. 616, 304 NW2d 455. See "Pushing the Boundaries of Eminent Domain," *Business Week* 5/4/81. p. 174.

[27] 104 S.Ct. 2321.

[28] 104 S.Ct. at 2330.

[29] *Id.* at 2329–30.

[30] City of Oakland v. Oakland Raiders, 32 Cal. 3d 60, 646 P.2d 835, 183 Cal. Rptr. 673 (1982), *cert. den.*, 30 June 1986. See Susan Crabtree, "Public Use in Eminent Domain: Are There Limits after *Oakland Raiders* and *Poletown?*" 20 *Calif. Western L. Rev.* 82, 88–90 (Spring 1984).

[31] See Gregory S. Alexander, "The Concept of Property in Private and Constitutional Law: The Ideology of the Scientific Turn in Legal Analysis," 82 *Columbia L. Rev.* 1545–99 (Dec. 1982). See also Margaret Jane Radin, "Property and Personhood," 34 *Stanford L. Rev.* 957–1015.

[32] See Flemming v. Nestor, 363 U.S. 603 (1960); Board of Regents v. Roth, 408 U.S. 564 (1972). Compare Pennsylvania Coal Co. v. Mahon, 260 U.S. 393 (1922).

[33] Hume, "Of the Original Social Contract," in *Social Contract*, ed. Ernest Barker, New York & London: Oxford Univ., 1960, p. 151; Locke, "An Essay concerning the True Original, Extent and End of Civil Government," in *Social Contract, op. cit.*, pp. 103–15.

[34] Rousseau, "The Social Contract," in *Social Contract, op. cit.*, p. 214.

[35] *Ibid.*, p. 216.

[36] *A Theory of Justice, op. cit.*, p. 197.

[37] Alan Riding, *Distant Neighbors: A Portrait of the Mexicans*, New York: Vintage, 1986, 1984, pp. 32–3, 51–2, 262, 420–1; Linda Schele and Mary Ellen Miller, *The Blood of Kings*, New York: Braziller, 1986. Special thanks to Sandy Hall, resident of Palenque.

[38] Noam Chomsky and Edward S. Herman, *The Washington Connection and Third World Fascism*, Boston: South End, 1979, pp. 129–204. See Also *ibid.*, pp. 205–17.

[39] Armand Mattelart, *Transnationals and the Third World: The Struggle for Culture*, tr. D. Buxton, South Hadley, MA: Bergin & Garvey, 1983, pp. 107–8.

[40] Maria Shao, "Why Taiwan's Doors Should Swing Open," *Business Week*, Aug. 3, 1987, p. 41; Douglas R. Sears, "Taiwan's Export Boom to U.S. Owes Much to American Firms," *Wall Street J.*, May 27, 1987, pp. 1, 12; "America's New-Wave Chip Firms," *ibid.*, p. 30; "Taiwan's Wealth Crisis: It's $53 Billion Cash Hoard is Economic Poison," *Business Week*, April 13, 1987, pp. 46–7.

[41] Frontline, PBS, June 2, 1987.

[42] Mattelart, *op. cit.*, p. 106.

[43] See, for example, Michael Novack, "Two Views on Helping Latin American Poor," *Indianapolis Star*, June 14, 1987.

[44] Mattelart, *op. cit.*, p. 111; Riding, *op. cit.*, pp. 417–20; "Mexico Looks Better and Better to Japan," *Business Week*, June 8, 1987. See also "U.S. Runaway Shops on the Mexican Border," *Nacla's Latin American and Empire Report* (New York) 9 no. 5 (July-Aug. 1975).

[45] Illustrative of current philosophical discussion of corporate responsibility are the following three works all published in Englewood Cliffs, NJ, by Prentice-Hall: Thomas Donaldson, *Corporation and Morality* (1982); Clarence Walton, ed., *The Ethics of Corporate Conduct* (1977); Patricia H. Werhane, *Persons, Rights and Corporations* (1985). See also Brent Fisse and Peter A. French, eds., *Corrigible Corporations and Unruly Law*, San Antonio: Trinity University, 1985.

[46] See "Shareholders Aren't Just Rolling Over Anymore," *Business Week*, April 27, 1987, p. 32; "The Battle for Corporate Control," *Business Week*, May 18, 1987, pp. 102–109.

[47] See Dinham and Hines, *op. cit.*; Richard J. Barnet and Ronald E. Muller, *Global Reach: The Power of the Multinational Corporations*, New York: Simon & Schuster, 1974, pp. 254–302.

[48] See, for example, the Marxist analysis in Elizabeth Fox Genovese and Eugene D. Genovese, *Fruits of Merchant Capital*, New York and Oxford: Oxford Univ., 1983; and the religious analysis in David Brion Davis, *The Problem of Slavery in Western Culture*, Ithaca: Cornell Univ., 1966.

[49] See "The End of Corporate Loyalty?" *Business Week*, Aug. 4, 1986, pp. 42–9.

[50] See John Gibbons, *Tenure and Toil*, Philadelphia: Lippincott, 1888, pp. 148–9, 183–91; Stanley Buder, *Pullman: An Experiment in Industrial Order and Community Planning, 1830–1930*, New York et al.: Oxford, 1967; Alfred Winslow Jones, *Life, Liberty, and Prosperity*, New York/London: Lippincott, 1941; Barbara Dinham and Colin Hines, *Agribusiness in Africa*, Trenton, NJ: Africa World Press, 1984.

STANLEY R. CARPENTER

WHAT TECHNOLOGIES TRANSFER:
THE CONTINGENT NATURE OF CULTURAL RESPONSES

INTRODUCTION

For all but the true believers the advocacy of technology transfer from the industrialized West to the "less developed countries" (LDCs) is starting to lack conviction. It is not denied that relationships of technological dependence are to be preferred over older colonial relationships. Nor has the thought been completely abandoned that technology transfer can accelerate the rate of diffusion of industrial technology to the LDCs. But events such as confirmed reports of massive destruction of the world's rain forests, believed to be crucial to the hydrologic cycle, by means of Western supplied equipment, or population problems exacerbated by "Green Revolution" measures, give pause to all but the technological pollyannas.

One creative reaction to the failings of technology transfer was that of the late E.F. Schumacher. Concern that LDCs often lacked the infrastructure required to sustain direct transplants of capital intensive Western technology led to his advocacy of "intermediate technology,"[1] Today the organization he started, Intermediate Technology Development Group, is at work in twenty or more countries.[2] Despite the positive contributions of this approach, its impact will likely remain limited. Intermediate technology, like its capital intensive counterparts, assumes the desirability of the delivery of technology from host to developing country. While the use of locals is encouraged, especially that of school teachers, the solutions tend to be hardware oriented. Schumacher's philosophy of technology, borrowing from both Catholicism and Buddhism, and encapsulated in his often reprinted "Buddhist Economics"[3], strikes one as highly suggestive and oracular, but lacking in substantive depth or systematic coherence.[4]

In the U.S., concerns about the shortcoming of technology transfer programs have been joined by growing self doubts that the technological practices employed at home, and heretofore the model for imitation and transfer, are sustainable. We have ourselves become a net debtor nation. Steel, machine tools, pharmaceuticals, chemicals, consumer

163

Edmund F. Byrne and Joseph C. Pitt (eds.), Technological Transformation:
Contextual and Conceptual Implications, 163–177.
© *1989 Kluwer Academic Publishers.*

electronics, memory chips, automobiles, textiles, electrical machinery are made elsewhere, often are of superior quality to our own. The work force lags in productivity behind Japan, Germany, France, Britain and Italy. The country whose creation and application of industrial technology has been most often admired and imitated finds itself possessing poorer air quality than many of the other industrial nations, fifteenth in life expectancy and eighteenth in infant mortality rates.[5] Is the loss of competitive position by the U.S. a mere perturbation in the ongoing and inevitable advancement of the technological "superculture?"[6] Is the relatively stable self-correcting mechanism of the market still at work as the technologies of the industrial countries are transferred, adopted and projected back into the world economy?

The role of technology is treated rather benignly when serious problems such as these are pinpointed. One finds problems couched in economic terms: adjustment of trade imbalances, reduction of domestic deficit spending, pollution surcharges, modifications of currency exchange rates, reductions of tariff barriers, forgiveness of LDC debts, etc. Despite the motor of technological innovation and diffusion impelling the process, technology is spared critical scrutiny. It is often touted as a panacea, seldom villain. With admiration and due respect to Jacques Ellul for having made the opposing case,[7] I believe his totalization of contemporary action within the "technological milieu" is hard to come to terms with. What seems such an all encompassing determinism produces little beyond despair.

While technology has tended to escape intensive criticism within the wider culture, there have been notable exceptions. Philosophical reflection on technology has occurred for a century among German engineers.[8] Major philosophers, notably Heidegger and Habermas, have profoundly and creatively analyzed technology. Additionally, at the risk of sounding self-congratulatory, it is gratifying that for over a decade now, the Society for Philosophy and Technology has provided a forum for Continental and Anglo-American exchanges among philosophers intent on increasing attention upon technology as a subject rich in philosophical issues.[9]

Along these lines the following discussion provides reasons as to why technology transfer is inevitably culture-laden. It continues a line of argument advanced by Dickson[10] to the effect that conceptions of technology are in one case misleading, when technology is invested with a transparent instrumentality and neutrality, and in a second case incomplete when technologies are assessed extrinsically in terms of, say, economic, political, or legal institutions.

This second approach will be augmented by a model of technology that treats it as a symbolic, intentional embodiment, as, in fact, a language of social action. Such a line of argument inextricably links technology and culture, but, more importantly, shows how modes of technological action reflexively condition the culture itself. So it is that technology creates a way of being in the world, or a world in which to be – in Winner's apt borrowing of Wittgenstein, a "form of life."[11] While technology is but one symbolic embodiment of culture, its pervasiveness – what Ellul and Heidegger both profoundly lament – provides an inescapable backdrop for the most serious of contemporary conversations. It is "well-formed" solutions, expressed in the syntax of Western technologies, by which aspirations of the LDCs are so frequently formulated. It is hoped that the following sheds some light on how this could be.

CULTURE-FREE TECHNOLOGY

My colleague Melvin Kranzberg, Callaway Chair of History of Technology, has proposed a law to the effect that "Technology is neither good nor bad, nor is it neutral!" While semi-humorous in form, an underlying fact is clear. It is increasingly difficult to treat technological phenomena as pure instrumentality. Yet a persistent refusal to invest technology with much significance reflects what Dickson calls one of the major myths of contemporary society.

The substance of the 'technology is politically neutral' myth is that the outcome of any particular act involving the application of an element of technology is determined entirely by the motives of the act. The technology utilized is held to be entirely independent of such motives.[12]

I have tried to suggest elsewhere that the reasons for the appeal of the idea that technology is neutral are crucially intertwined with the beginnings of modern Western philosophy.[13] Indeed, the same subject/object dichotomy generated by Descartes' ontological dualism legitimates the notion of the disengaged self. Human subjects distance themselves from the world by making it an object of scrutiny. Human intentions are considered separable from the physical space of their realization. Charles

Taylor argues at length that this model of disengagement results in a definition of the self as disembodied, punctual, and atomistic.[14] Engagement, such as it is, must be characterized as utilitarian, instrumentalist, and remote. Forms of symbolic expression, whether ideational, as in the case of natural or artificial languages, or materially realized in the form of tools, or instruments, are transparent conduits of human intentions, and are thus themselves morally neutral.

The Cartesian subject/object split, perpetuated in both rationalist and empiricist foundationalism, achieves its clearest articulation in positivistic interpretations of science. Matters of human intention are separable from matters of fact. It is the latter, furthermore, of which objective knowledge is constituted. Consequently, while intentions may be appropriate subjects of moral scrutiny, objective knowledge is value-free and, over time, increasingly in touch with what is real.

The tendency among scientific practitioners remains strong to distinguish sharply between the *products* of science: journal articles, and increasingly, prepublication drafts privately circulated; and the varied and often idiosyncratic *processes* that led to their production. It is the former on which academic advancement is mainly based.

If this discredited dichotomy between the contexts of justification and discovery nevertheless barely manages to survive, its technological counterpart appears to be more vigorous and youthful than ever. The fashioning of machines, instruments, and tools is strictly a technical process. Design constraints are dictated by "technical efficiency,"[15] a norm reflecting calculability, objectivity and a heartless independence from human desires. The criterion of efficiency is particularly powerful if expressed in mathematical form, with the demonstration of an increased input/output ratio possessing near self-certifying authority. When such influence, derived solely from technical calculation, is used to legitimate technical and social change, it is best characterized as scientistic. "The basic message of scientism," Dickson observes, "is that an apparently 'scientific' approach to any problem or situation is both necessary and sufficient to indicate how its objective, politically neutral resolution can be achieved."[16]

Rorty has popularized the metaphor of the "mirror of nature" as applying to foundationalist epistemologies,[17] but the analogy is strikingly extendable to the technological embodiments of foundational scientific knowledge. The invented artifact mirrors the objectively defined limit of technological possibility itself. The dictates of applied science are transmitted through the highly trained engineer and given objective form.

An opposing position to the "mirror of nature" view of science, and by extension its applied embodiment in technology, rejects the scientistic description of technological design. Joseph Margolis' recent thinking is highly suggestive in this regard. By arguing against the cognitive transparency of nature and refusing to accept the foundationalist articulation of the subject/object split, he defends a "minimal realism" while at the same time continuing to maintain that scientific achievement is unavoidably culture-laden and historically implicated.[18] This non-foundationalist and non-essentialistic construal of human culture, including science and technology, makes awareness, understanding, and inventiveness respective instances of order that is devised more than discovered, by human societies which are always historically and culturally situated. We will return to Margolis' position below.

If the position that technology neutrally mirrors the range of material possibility is indefensibly naive, as I believe it is, and if the technological act is seen as deeply situated and culturally embedded, both temporally and spatially, our conception of technology transfer is complicated. Technology transfer becomes clearly an instance of culture transfer. This point is not missed on the elite youth of the LDCs sent to the West for advanced education, who soon realize that what they will return home with is more than a bag of techniques. This fact ought to give us educators pause. The transfer of technologies from cultures possessing a tradition of democratic liberalism to those where it is lacking involves more than the passing along of a politically neutral tool. It amounts to the settling of political questions in democratically questionable ways.

A clear measure of the originality of E.F. Schumacher is seen in the evolution of his thinking concerning technology transfer. His twenty years as top economist and head of planning of the British Coal Board provided ample opportunity for prescribing remedies for the economic ills of the post-colonial countries. His appointment in 1955 as temporary Economic Advisor to the government of Burma produced a change in his conception of technology transfer. As related by his wife

The temporary Economic Advisor was impressed, not only by the homogeneous nature of the community, but also by its non-violent character. Clearly, such a people needed a non-violent form of technology for their material development – one which would not disturb their natural social cohesiveness, or clash with their religious beliefs.[19]

Schumacher himself wrote:

When I came to ask myself this question, "What would be the appropriate technology for rural India or rural Latin America or maybe the city slums?" I came [he says] to a very simple provisional answer. That technology would indeed be really more intelligent, efficient, scientific, if you like, than the very low level technology employed there, which kept them poor. But it should be very, very much simpler, very much cheaper, very much easier to maintain than the highly sophisticated technology of the modern West. In other words, it would be an *intermediate technology*, somewhere in between.[20]

Schumacher's first hand experiences which led him to revise his ideas about development is consistent with a rejection of the idea of culturally neutral technology transfer. An analysis of technology should be undertaken along different lines. One such approach, to which we now turn, analyzes technology categorically, with each category constituting a more or less separable social institution. Rather than regarding technology as an instrumentally neutral conveyance of human intention, this position somewhat more directly implicates technology in the imposition of economic, scientific, and political order upon human societies.

TECHNOLOGY INSTITUTIONALIZED

For one and one half decades the U.S. Office of Technology Assessment (OTA) has provided the members of the Congress, along with interested citizens, its best estimates of the impacts of newly envisaged or modified technologies.[21] Requests for these analyses come from the Congress on an ad hoc basis. The approach is case specific although assessments are often quite comprehensive.[22]

The procedure for performing a technology assessment study (TA) amounts to an analysis of a particular technology along modal impact categories. My colleagues and I once suggested an acronym – EPISTLE to assist such analyses.[23] One thereby devises an impact checklist according to Economic, Political, Institutional, Social, Technological, Legal, and Environmental dimensions. There is no implied order of importance to the list, although the presence of economics as a lead analysis is consistent with characteristics pride of place. With all its shortcomings[24] TA does reflect the original consensus of the 1960s that forestalling technological disasters as well as spotting fortuitous opportunities requires focused effort and targeted analyses in place of blind faith in the self-correcting tendencies of market-driven technological change.

The sociological theory of institutions provides a related but deeper approach to understanding technology.[25] In order for complex patterns of human behavior to acquire institutional status they must organize and focus action in ways that are uniquely valued for species survival. Thus, kinship, religion, politics, the military are typically identified as paradigmatic social institutions. Social institutions compete with one another, an event both inevitable and desirable and one comprising what Robert Nisbet calls "ethical history."[26]

The Enlightenment is noteworthy for having led to the emergence of a new social institution, so argue both Robert Heilbroner[27] and Karl Polanyi. In Polanyi's words:

(N)o society can exist without a system of some kind which ensures order in the production and distribution of goods. But this does not imply the existence of separate institutions; normally the economic order is merely a function of the social, in which it is contained. Neither under tribal nor feudal, nor mercantile conditions was there . . . a separate economic system in society. Nineteenth century society, in which economic activity was isolated and imputed to a distinctive economic motive, was, indeed, a singular departure.[28]

Economics, more than other social institutions, illustrates the incompleteness of an institutional analysis of technology. While technology is undeniably a major force for change it is cast in an ancillary role, held in check by the economic "bottom line." It is the dependent variable vis-à-vis economic viability. Schumacher makes the point poignantly.

In the current vocabulary of condemnation there are few words as final and conclusive as the word 'uneconomic'. If an activity has been branded as uneconomic, its right to existence is not merely questioned but energetically denied. Anything that is found to be an impediment to economic growth is a shameful thing, and if people cling to it, they are thought of as either saboteurs or fools. Call a thing immoral or ugly, soul-destroying or a degradation of man, a peril to the peace of the world or to the well-being of future generations; as long as you have not shown it to be 'uneconomic' you have not really questioned its right to exist, grow, and prosper.[29]

When short term concerns about the bottom line are effectively isolated from long term questions of species survival or resource preservation, as is nearly always the case when economic models are utilized, it should be clear that our analytical tools need supplementing. What is required is an approach that places technology closer to the core of the analysis, not at its surface.

Robert Nisbet expands the theory of social institutions to include technology in an attempt to provide such an approach. He writes:

In every relevant respect, technology is today as fully and distinguishably an institution as law, religion, or kinship. It is neither more nor less "material" than other institutions. There is no more reason for limiting technology, as a concept, to the machines and tools it employs than there is for limiting the family to housing, law to courtrooms, or religion to church buildings.[30]

Drawing out the implications of this claim would run somewhat as follows:

Technology as a social institution contains its own form of total commitment. It possesses its own inner logic, values and goals. It justifies its activities and enters into competition and conflict with other institutions according to canons of rationality which it itself defines. It raises the human capacity to manipulate and reorder the physical environment to an end in itself. Just as economics earlier gained autonomy by turning natural surroundings into "land," and human beings into "labor," technology has transformed human mobility into having a car, communication into having a telephone (and soon, probably, having a computer), and nourishment needs into refrigerators, stoves, and supermarkets.[31]

While I still regard this contention that technology is to be understood as a social institution as possessing significant explanatory power, I wish in the final section to provide an augmentation, to the effect that technology should also be regarded as a language, namely, a language of social action. Technology is to be interpreted as more that just a major embodiment of culture. To the extent that we see culture as historically contingent, we may properly regard technology as delimiting social action and understanding as well as creating the categories by which cultural experience is itself rendered intelligible.

TECHNOLOGY AS LANGUAGE

The claim that technology is a language, indeed one of social action, requires elaboration. A general conception of language needs stipulation. It is no overstatement to view a preoccupation with intentionality as central to twentieth century Western philosophy. Viewpoints may be grouped into two opposing categories. One considers intentionality of thought to be prior to that of speech and accepts the existence of independent "atoms" of meaning. Frege, Russell, Carnap and the Wittgenstein of the *Tractatus*, for example, support this position. The other claims intentionality of speech to be prior to that of thought and regards society as the ultimate meaning giver. Dewey, Heidegger, the later Wittgenstein, Sellers, Kuhn, and Rorty are some of the proponents of this position. While the first position suggests that meanings can be specified independently of history, the second holds meanings to be embedded within the cultural-historical context.[32]

According to Charles Taylor, the first position – what he calls the "designative theory of language" – came into its own with Descartes, Bacon and Hobbes. On this view words are the basic instruments of meaning. "(L)anguage is a set of instruments which lie as it were transparently to hand and which can be used to marshall ideas . . . "[33] When technology is understood as functioning in a purely instrumental fashion, as was discussed above, it appears to manifest the same transparency as do other symbolic forms. This misconstrual of the symbolic role of technology leads to the erroneous conclusion that technology is itself neutral.

The second conception of language Taylor calls "expressive" and finds its early defense in late eighteenth century and early Romantic thought. Whereas the Romantics took expression to be *self*-expression, Heidegger, he notes, took it to be mainly about the world. Expressivism sees language as making manifest a way of being in the world. "(T)he use of a word is always in a context, a matrix. . . . words we use now only have sense through their place in the whole web . . . "[34] Language is thus best understood holistically rather than atomistically; it is constitutive of awareness, rather than merely its articulation.

What then does language come to be on this view? A pattern of activity, by which we express/realize a certain way of being in the world, that of reflective awareness, but a pattern which can only be deployed against a background which we can never fully dominate; and yet a background that we are never fully dominated by, because we are constantly reshaping it.[35]

This expressive and holistic conception of language, much more than the designative viewpoint, identifies language origination and use with human creativity. Taylor notes the impossibility of drawing a boundary around prose language and distinguishing it from poetry, music, art, dance. Thus it is language as symbolic expressiveness, manifest in a variety of modes, by means of which we acquire reflective awareness, interact with our surroundings, and establish human relationships.

David Dickson's assertion that technology is a language of social action does not, in my view, do damage to Taylor's expressivist model, and it does focus the issue in the direction I find suggestive.

(I)n the way that language is experienced through the spoken or written word, which therefore become the 'objectifications' of social ideas and concepts, so . . . technology represents in the individual machines of which it is composed the 'objectification' of social action.[36]

What is it that the language objectified in technology expresses? Dickson observes:

(T)he articulation of a particular aspect of technology as a social institution already coincides with the way that it is intended the technology should be used. . . . (A) society's technology, when viewed as a social institution rather than a heterogeneous collection of machines and tools is structured in a way that coincides with its dominant modes of action and interaction.[37]

Is it possible to apply the expressive conception of language to technology, as Dickson does, without bringing along all and only Marxist presuppositions? Can technology be meaningfully considered as symbolically expressive of something other than existing power relationships? May not the point be intelligibly made that technological actions, artifacts, processes, organization, etc., express a way of taking up and being in the world that is not solely a reflection of control and allocation patterns of resources and tools?

Joseph Margolis's "minimal realism" with its elucidation of the "praxical" provides an affirmative answer. Using the term 'praxical' to refer to " . . . this indissoluble linkage between purposive activity and the work of man and what might be called the legibility of nature (including a fortiori, human culture)" he provides an analysis which he claims is neutral to the conceptual quarrels of Marx, Heidegger, Lukacs, Adorno, Althusser, Dewey, and Habermas.[38] What the position is not neutral about is its rejection of foundational forms of epistemological realism and essentialistic interpretations of the laws of nature. "(O)ur very understanding of the world is seen to be a function of the contingent, varied, and clearly transient forms of human intervention."[39] In the same passage the terms 'praxical' and 'technological' are used synonymously.

Margolis defends a position of non-reductive materialism in which cultural entities – persons, works of art, machines, parts of language – emerge within the praxical interchange.[40] Cultural entities are characterized as intentional. It is not stretching the point to treat technology straightforwardly as a language in this connection, namely, an intentional embodiment of the praxical. In place of the myth of technological neutrality, Margolis locates technology in the intersection of ideology and practical knowledge.

Technology is practical knowledge dominated by the realities of social existence among humans, that is, of ideological objectives affecting all means-ends sequences of practical knowledge. Such objectives cannot be simply characterized as efficiency or effectiveness, since those notions are bound to be considered in abstraction from social realities.[41]

Margolis claims that neither practical knowledge nor current realities of social existence are to be regarded as fixed, foundational, or inevitable. While it is true that practical knowledge is currently dominated by applied science, he shows elsewhere that science itself is to be construed as the work and achievement of historical communities, thus historicistically.[42] Similarly, the "current realities of social existence" should not be taken as determining the correct ideology, as Marxists would assert. The relationship between current practical knowledge and current ideology, that together comprise the praxical, is thus dialectical yet contingent.

Without offering us foundational certainty, Margolis both underscores the intentionality with which each technological act is freighted, and also provides relief from the determinism that Marx, and Ellul, for that matter, invest in their conceptions of technological change. He observes:

(I)t is precisely man's transcendental grasp of the technological grounding of his own sustained survival and cognitive achievement that most plausibly insures his sense of the structure of the real world and of his capacity to discern it. . . . The pivotal issue remains how to recover transcendental reflection under the constraints of a radically historicized pragmatism. The answer, quite simply, lies with the grasping the primacy of social *praxis* and, through that, the inescapably technological orientation of our reflection on the nature of reality. . . . (T)he technological represents pragmatism's best clue linking the tacit and cognitively explicit features of whatever realism we can justifiably ascribe to our science and effective routines of life.[43]

This way of thinking offers rich insights into the phenomenon of technology transfer. What technologies transfer, as Margolis shows, includes both ideology and practical knowledge. Since technology constitutes a major intentional embodiment of culture, its transfer cannot occur solely in instrumental ways but instead carries along donor country ideology. However, given the dialectical nature of technology, with both ideology and practical knowledge historically situated, what is transferred is itself transformed and reflected in the process. This is so both from pressures supplied by the recipient country and from threats to species survival gradually dawning on donor cultures. Thus, in one case, Schumacher's prescriptions for "intermediate technology" applied to the LDCs reflexively produced "alternative technology" movements in the industrial countries. In another, what Margolis calls a tacit species propensity to promote survival[44] increasingly calls into question conventional economic boundaries between the internal and external costs of technological development. In the process the pace of negotiation between technological forces and the broader culture quicken. New licenses as well as new proscriptions result. Because this dialectical process is driven by a complex and evolving view of species survival, it is not grounded in a fixed determinism of material conditions. A "soft determinism" better describes the process.

The technologies which transfer thus carry with them cultural syntax and implicit messages concerning acceptable patterns of action. But as this language evolves from emerging survival pressures, both form and content change.

CONCLUSION

We have argued that understanding technology transfer is crucially dependent on a conception of technology that complements a fairly obvious instrumental model with broader cultural factors. Beyond the genuinely misleading view that technology can transfer in culturally neutral ways, we have acknowledged the importance of a modal analysis of technology along political, economic, religious, etc., lines. This approach, however, needs augmentation which is accomplished in part by considering technology in sociological categories as an institution, complete with inner logic, values, goals, and ways of being in the world. Understanding technology as a social institution clarifies the culture-ladenness of technology generation and transfer.

The third conception of technology goes beyond this institutional conception to consider technology as itself a language – the language of social action. Here Margolis' rich theory of culture locates our categories of understanding themselves in the technological. Self-awareness of our being in the world reflect technological categories and thus provide our definitive orientation. Yet, this fact is itself historically contingent. We speak with our technologies in the act of transferring them, yet the syntax changes episodically.

It is currently fashionable to claim that we are witnessing the end of individualism in the West, at least the extreme form represented by the U.S.[45] It is argued that the sharp separation between the civic and business cultures that have characterized U.S. industrial development, may now be working to our disadvantage as a nation.[46] If Margolis' optimism regarding the tacit survival wisdom of our species is borne out, we may hope and expect new blueprints for the future, one less atomistic in their basis and less benign toward the independence of the business culture. One can hope, as he does, for "relatively nontendentious directives for technology in general"[47] that reflect a prudential species interest about such matters as population control, resource renewal, flora and fauna preservation, pollution reduction. George Grant observes that "the coming to be of technology has required changes in what we think is good, what we think good is, how we conceive sanity and madness, justice and injustice, rationality and irrationality, beauty and ugliness."[48] The point is to make every effort to foster conversations that keep these dichotomies open for negotiation in ways consistent with prudential wisdom.

NOTES

[1] E.F. Schumacher (1973) *Small Is Beautiful: Economics As If People Mattered*, New York: Harper Torchbooks: (1977) *A Guide for the Perplexed*. New York: Harper Colophon: with Peter N. Gillingham (1979) *Good Work*. Harper Colophon.
[2] George McRobie (1981) *Small Is Possible*. New York: Harper & Row, Publishers.
[3] E.F. Schumacher (1973), note 1, pp. 50–58.
[4] For an extended critique of intermediate technology in the context of an assessment of the alternative technology (AT) movement see Stanley R. Carpenter, "A Conversation Concerning Technology: The Appropriate Technology Movement." in *Philosophy and Technology*, Vol. 4., Paul T. Durbin, Ed., Dordrecht: D. Reidel, pp. 87–105.
[5] Robert B. Reich (1983) *The American Frontier*, New York: Penguin Books, pp. 117–18.
[6] According to Kenneth Boulding, the superculture is epitomized by the international similarity of airports, throughways, skyscrapers, hybrid corn and artificial fertilizers, birth control, and universities. (1969) "The Emerging Superculture: The Interplay of Technology and Values," in Kurt Baier & Nicholas Rescher, eds., *Values and the Future*. New York: The Free Press, pp. 336–50.
[7] Jacques Ellul (1967) *The Technological Society*. Trans. by J. Wilkinson. New York: Vintage Books.
[8] For a summary discussion of the German philosophy of technology movement with its close ties to German engineering, as well as annotated bibliography of other investigations of this movement see Friedrich Rapp (1981) *Analytic Philosophy of Technology* Trans. S. Carpenter & T. Langenbruch. Volume 63, Robert S. Cohen and Marx W. Wartofsky, Series eds., *Boston Studies in the Philosophy of Science*. Dordrecht: D. Reidel., pp. 4–16.
[9] Cf., esp. Paul T. Durbin (1977) "Are There Interesting Philosophical Issues in Technology as Distinct from Science?", pp. 139–52, and Mario Bunge (1977) "The Philosophical

Richness of Technology," pp. 153–72; both in *PSA 1976*, Vol. 2. East Lansing, MI: Philosophy of Science Association.

[10] David Dickson (1974) *The Politics of Alternative Technology*. New York: Universe Books, esp. 174–204.

[11] This is a central theme of both of Winner's books: (1986) *The Whale and the Reactor*. Chicago: University of Chicago Press, esp. pp. 3–18; and his earlier (1977) *Autonomous Technology: Technics-out-of Control as a Theme in Political Thought*. Cambridge, Ma: MIT Press.

[12] Dickson, note 10, p. 183.

[13] Stanley R. Carpenter, "A Conversation Concerning Technology: The 'Appropriate' Technology Move," to appear in *Philosophy and Technology*, Vol. 4., Paul T. Durbin, Ed., Dordrecht: D. Reidel (forthcoming).

[14] Charles Taylor (1987) "Overcoming Epistemology," in Kenneth Baynes *et al.* (eds), *Philosophy: End or Transformation?* Cambridge, MA: MIT Press, pp. 464–88.

[15] Cf. Stanley R. Carpenter (1983) "Alternative Technology and the Norm of Efficiency," in Paul T. Durbin, (ed), *Research in Philosophy & Technology*. Vol. 6, pp. 65–76.

[16] Dickson, note 10, pp. 185, 86.

[17] Richard Rorty (1979) *Philosophy and the Mirror of Nature*. Princeton: Princeton University Press, and (1982) *Consequences of Pragmatism: Essays 1972–1980*. Minneapolis, MN: University of Minnesota Press.

[18] Joseph Margolis (1986) *Pragmatism without Foundations: Reconciling Realism and Relativism*. Oxford: Basil Blackwell Inc. It is a pleasure to note that this volume is the first in an envisaged trilogy. Soon to follow are "Science without Unity: Reconciling the Human and Natural Sciences" and "Texts without Referents: Reconciling Science and Narrative."

[19] Verena Schumacher in Forward to George McRobie (1981), note 2, p. xi.

[20] *Ibid.*, p. 6

[21] The OTA origination law was PL 92–484, signed October 13, 1972.

[22] An indicative sampling: "Life Sustaining Technologies and the Elderly" (7/87); "Technology Transfer to China" (7/87); "Wastes in Marine Environments" (5/87); "Technologies to Maintain Biological Diversity" (3/87); "Technologies for Detecting Heritable Mutations in Human Beings" (9/86); "Transportation of Hazardous Material" (7/86); "Intellectual Property Rights in an Age of Electronics and Information" (4/86); "Alternatives to Animal Use in Research, Testing, and Education" (2/86); "Electronic Surveillance and Civil Liberties" (10/85). Source: "OTA Report Briefs," OTA, U.S. Congress, Washington, D.C.

[23] A. Porter, F. Rossini, S. Carpenter, T. Roper (1980) *A Guidebook for Technology Assessment and Impact Analysis*. New York: North Holland.

[24] Cf., Stanley R. Carpenter (1983) "Technoaxiology: Appropriate Norms for Technology Assessment," in Paul T. Durbin and Friedrich Rapp, eds., *Philosophy and Technology*. Robert S. Cohen and Marx W. Wartofsky, Series eds., *Boston Studies in the Philosophy of Science*. Volume 80. pp. 115–36.

[25] Cf., Stanley R. Carpenter (1984) "Redrawing the Bottom Line: The Optional Character of Technical Design Norms," *Technology in Society*. Vol. 6, pp. 329–40.

[26] Robert Nisbet (1968) "The Impact of Technology on Ethical Decision-Making," in *Tradition and Revolt*. New York: Random House, p. 189.

[27] Robert Heilbroner (1975) *The Making of Economic Society*, 5th ed. Englewood Cliffs, NJ: Prentice-Hall, pp. 47–68.

[28] Karl Polanyi (1944) *The Great Transformation*. New York: Rinehart. p. 71.

[29] E.F. Schumacher (1973), note 1, pp. 39, 40.

[30] Robert Nisbet (1968), note 26, p. 185. As a point of comparison, Jacques Ellul's original *La Technique ou l'enjeu du siecle*, appeared in 1954, Paris: Armand Colin, and appeared in English in 1964 as *The Technological Society*. Trans. by J. Wilkinson, New York: Knopf.

[31] Stanley R. Carpenter (1983), note 15, pp. 68, 69.

[32] A summary of the dichotomous positions regarding "intentionality" in twentieth century Western philosophy is provided in *APA Newsletter on the Teaching of Philosophy*, vol. 2, #3 (1981, Spring), p. 2.

[33] Charles Taylor (1985) *Human Agency and Language: Philosophical Papers I*. Cambridge: Cambridge University Press, p. 218.

[34] *Ibid.*, p. 231. Taylor cites as his exemplars of this position the nineteenth century "romantic" thought of Herder and Humboldt.

[35] *Ibid.*, p. 232.

[36] David Dickson, note 10, p. 176.

[37] *Ibid.*, pp. 178, 80.

[38] Joseph Margolis (1988) "Information, Artificial Intelligence and the Praxical," paper presented at the International Philosophical Conference on Information Technology and Computers. Tarrytown, NY, September 1983, p. 1. (mimeo).

[39] *Ibid.*

[40] Joseph Margolis (1978) "Culture and Technology," in Paul T. Durbin, ed., *Research in Philosophy & Technology, Vol. 1*. Greenwich, CT: JAI Press, Inc., p. 32.

[41] *Ibid.*, pp. 34, 35.

[42] Joseph Margolis (1986), note 18, p. 116.

[43] Joseph Margolis (1983) "Pragmatism, Transcendental Arguments, and the Technological" in Paul T. Durbin and Friedrich Rapp, (eds), *Philosophy and Technology*. Robert S. Cohen and Marx W. Wartofsky, Series (eds), *Boston Studies in the Philosophy of Science*, Volume 80, pp. 302, 3.

[44] *Ibid.*, p. 299.

[45] Cf. the enormously popular Robert N. Bellah *et al.* (1985), *Habits of the Heart: Individualism and Commitment in American Life*. Berkeley: Univ. of California Press.

[46] Robert Reich, note 5, pp. 3–21.

[47] Joseph Margolis (1978), note 41, p. 35.

[48] George P. Grant (1986) *Technology and Justice*. South Bend, IN: Univ. of Notre Dame Press, p.32.

BERNARD DENOUDEN

TRANSFERRED AND TRANSFORMED TECHNOLOGY:
THE C.R.S. THRESHER/WINNOWER

In the brief presentation that follows, I will argue by example that a technology can be creatively transferred if it is appropriately transformed and
designed to fit into the culture and economy of the people and country for
which it is intended. I will also argue that this kind of technological transfer
can foster independence rather than dependence and that it can generate a
degree of economic freedom and not be part of a mechanism that
fosters inseverable economic ties to a donor country. To achieve these
goals or perhaps ideals, a set of very clear criteria must be created for
each project or technological phenomenon. My argument takes the
form of a case study of the Catholic Relief Service thresher-winnower
and the criteria that were developed for its creation, production, implementation and evaluation.[1] The source of material for this study was
an examination of the research, planning, design and evaluation process
of the machine and a field visit to Egypt and to the factory where this
thresher is produced. I also accompanied a maintenance team in the
delivery of a unit that was put into production that same day in the
agricultural lands of the Nile Delta. In addition, I spent one month
reviewing and evaluating agricultural and development projects of
various kinds in Egypt.[2]

An interesting example of appropriate and transferrable technology is
the C.R.S. multi-cropper thresher-winnower. It was developed in
Egypt by Catholic Relief Services with the support of funding from the
Ford Foundation. It was designed after a thorough needs survey which
made evident a compelling requirement for the improvement of threshing techniques and the reduction of labor costs thereto related. Ironically, in spite of Egypt's growing population and dense concentration of
people on agricultural lands, farmers consistently commented when
surveyed that there was a perennial shortage of labor during harvest
time. Labor cost for harvest in the Nile Delta averages from 5 to 6
Egyptian pounds per day for each hiree, which is equivalent to 3 or 4
American dollars. Threshing with a flail and winnowing by throwing grain
in the air is a backbreaking, boring and exhausting task. It is one of the

179

Edmund F. Byrne and Joseph C. Pitt (eds.), Technological Transformation:
Contextual and Conceptual Implications, 179–184.
© *1989 Kluwer Academic Publishers.*

most important activities in the agricultural process because without a clean and well-threshed harvest, all the plowing, planting, cultivating, hand weeding and irrigation are wasted.

The goal of the C.R.S. project was to save labor, but not to radically displace agricultural workers.[3] The grains in Egypt are, for a greater part, still cut and bound by hand. They are hauled to the thresher by traditional means of conveyance, such as a donkey cart, camel, or when available, tractors and wagons from the Egyptian cooperatives. Three individuals are required to operate the thresher. Workers still must be hired to cut or reap the grain. However, when the exhausting, mindless task of flailing the grain would be the next stage in the process of harvest, a small, simply designed threshing machine provides an efficient, if not exhilarating, finish to this activity. The thresher is 98% efficient; that is, after threshing only 2% of the grain can be found in the straw or stalks. It is small, such that it can be moved about by 4 or 5 men or pulled easily from farm to farm by a donkey. Remarkably, it can thresh and winnow "all major cereal and legume crops in Africa and the Middle East with two easily made adjustments".[4]

The reason the thresher is successful is that from needs study to design and development through field testing and refinement, it was carefully synchronized into the culture and economy for which it was intended. Motor powered threshers were, as you know, developed in technologically advanced countries. This technology was transferred, transformed and redesigned in Egypt for the agricultural infrastructure of developing countries. "Its goal was to help small farmers counter critical labor shortages at peak harvest periods. This bottleneck had caused significant crop losses and reduced farm income."[5] The thresher is manufactured in Egypt. It can be repaired in any rural blacksmith or machine shop with the simplest of tools. Its goal is to foster independence rather than potentially exploitive dependence. It does not need Western technology, with the exception of the small diesel engine that is extremely efficient. This engine runs at full capacity on one liter of fuel an hour. The engine that is presently used after elaborate research is a 10 horsepower German diesel that has an expected life span before overhaul of some 3,000 hours. Some engines of this type have lasted 7,000 hours before major repair. This, of course, presumes reasonable maintenance, and an essential dimension of this project is maintenance training and consistent follow-up visits to insure that everything is working properly. Central to the follow-up visits is making certain that the user farmers understand the machine's functions and maintenance requirements. This machine is known as the multi-cropper and was designed to meet the following criteria:

1. It must save labor;
2. It must be predominantly locally manufactured;
3. It must be simple;
4. It must be priced economically;
5. It must increase food production by reducing post-harvest losses;
6. It must generate additional farm income;
7. It must be socially acceptable.[6]

The C.R.S. thresher clearly meets the criteria that were established prior to its development and implementation. The machine quite obviously does save labor. Instead of a team of 9 to 10 laborers working for days to harvest the crop from 1 acre or fedan, this same task can be accomplished by 3 workers in approximately 3 hours. The result is a clean grain stored in sacks and ready for storage in a relatively cool and dry room. The machine, as I asserted previously, is locally manufactured. The factory where it is produced is located on the outskirts of Cairo. Many parts are made in this small and well-furnished facility. This factory has now expanded to produce mowing machines of a comparably simple design which can be used to cut clover or grain. This mower is not a reaper-binder, because in the judgement of the designers and planners a knotter too frequently malfunctions and thus a machine which would be too difficult to maintain would, in all likelihood, serve to be a source of frustration and would be abandoned. There is also a concern in the design of all the machines produced in this factory that a need is met without an imbalanced and inappropriate displacement of labor, farmers and laborers are still required to work through the field's binding sheaves individually with an ancient knot that you will find in Africa, Asia and in the histories of pre-industrial Europe. A large or even middle-sized combine directly imported from the U.S., Europe or the Soviet Union, whether it is self-propelled or drawn by a tractor, would not be appropriate to the economic needs of the Nile Delta or most other developing countries. In the Nile Delta it would severely compact the soft irrigated soil or be bogged down much of the time. However, the basic design element of the combined process of threshing and winnowing is transferable. The scale of this thresher is appropriate because of clear planning criteria and sound engineering practice. It fits the environment for which it was intended.[7]

The multi-cropper is simple. By opening the hood that covers the threshing drum one can readily see how it functions. It employs gravity feed after the grain is threshed. After winnowing, the grain is augered back up to bag level. "The machine can be simply set to thresh and winnow more than one dozen crops by exchanging a pulley drive belt and fitting appropriate screens for specific crops."[8] Since the design is not complex, repair is relatively easy to implement. The interrelated parts work leisurely together and the need for replacement parts is minimal. Pricing or the purchase cost of the thresher is extremely important. To create appropriate technology and then make it accessible only by gifts can also foster dependency. In this project the farmers must pay for the machine. The cost at present is approaching $3,000 per unit. The farmer must come up with a down payment and the remainder is payable by loan. The machine is income generating. Since the owner's harvest time is cut dramatically, he can hire out to other farmers at a savings to them and a profit to himself. A hardworking farmer can earn up to $3,300 in one year through harvesting crops other than his own. Planting and harvest in Egypt go on all year round because of favorable climate and irrigation. Two or three crops are often planted on each plot every year. This figure of $3,300 is almost 6 times the average annual income in rural Egypt.[9] The freedom that additional income can purchase is dramatic. Diet can be diversified and improved, the education of children is possible and investment in the improvement of other farming and living practices can be accomplished.[10]

The thresher-winnower does reduce post-harvest losses. Efficient processing of crops frees family members from spending days protecting harvest against insects, birds, rodents, fire, and theft while the crops wait to be processed for storage. Often crops have to be guarded around the clock to be protected against these contingencies. A quick, efficient threshing protects the quality of the grain and straw. Straw is used for animal fodder in Egypt, not merely for bedding. Animals are still the primary source of power in rural Egypt and thus a high quality fodder that hasn't molded or deteriorated is extremely important. The multi-cropper is clearly socially acceptable. It is most frequently used near the houses and storage areas of farm families. Women and children provide much of the farm labor in Egypt and, thus, to thresh and winnow near the house is compatible with the rather restricted sphere of environment that is socially acceptable for rural women in this country. I interviewed a number of individuals concerning the difference a thresher such as this would make, and all commented that it would save time, earn them additional income, and give them time to work at other activities. The economic and cultural disturbance that the presence of this thresher causes is gentle and creative. It fits a clear set of needs and provides for a potentially greater economic freedom for the owner and his/her neighbors. In my judgment, based on a review of the planning and evaluation documents, a visit to the factory where it is produced, and accompanying a delivery and set-up of this machine, the C.R.S. thresher-winnower is an excellent example of culturally transferable appropriate technology. What is essential is that the technology is not only transferred, but transformed into a logic that functions intelligibly and creatively within the intended culture. The machine is a success not merely because the West makes such wonderful machines. It has succeeded because some designs which evolved in the West were altered to creatively synchronize with the needs and agricultural infrastructure of developing countries.[11]

The C.R.S. thresher has stood up well to many evaluation studies and forms of field testing. The greatest promoters of it now are the farmers themselves. It has lived up to its own criteria for design and development and is being slowly adapted and improved as it functions in developing countries. Appropriate technology is not only an important concept to articulate and examine. It is a powerfully compelling and constructive phenomenon where it is translated into operation and production.

NOTES

[1] On the issue of criteria for appropriate technology, please see *Towards Global Action for Appropriate Technology*, edited by A.S. Bhalla, (New York: Pergamon Press, 1971), especially Chap. 1, "Appropriate Technology Criteria".

[2] This field study was made in January 1987. I reviewed projects sponsored by CARE, the United Nations, World Food Program, Catholic Relief Services, Environment Quality International and the Egyptian Government.

[3] For an excellent discussion of technology and displacement, see John David and Alan Ballard's *As Though People Mattered: A Prospect for Britain*, (London: Intermediate Technology Publications, 1986). Although this work focuses primarily on Britain, many interesting analogies can be drawn from it. In particular, Chapter 4, "Technology Choices" has constructs and analysis that are easily transferable to other environments and economies.

[4] "The C.R.S. Multi-Crop Thresher-Winnower: Development Through Small-Scale Agricultural Machinery." (A brochure published by C.R.S.), p 1.

[5] *Ibid.*

[6] *Ibid.*, p. 2.

[7] For a general discussion of the adaptation of technology and the role of donors in the choice of appropriate technology, see *Choice and Adaptation of Technology in Developing Countries: An Overview of Major Policy Issues.* (Development Centre of the Organization for Economic Cooperation and Development, Paris, 1974).

[8] C.R.S. brochure, page 1.

[9] *Ibid.*, p. 5.

[10] Synchronizing traditional activity with new technologies is of obvious importance. For a general study of this, see *Blending of New Technologies with Traditional Activities.* (Geneva: International Labor Org., First Session, Advisory Committee on Technology, April 15–19, 1985).

[11] For a general discussion of appropriate technology with a series of well-developed case studies, see *Experiences in Appropriate Technology*, edited by Robert J. Mitchell, (Ottawa: The Canadian Hunger Foundation, 1980.) For a comprehensive quantitative analysis of: 1) the growth of appropriate technology organizations; 2) the growth of appropriate technology activities; 3) the types of activities of these organizations; 4) the funding and staffing of appropriate technology organizations; 5) the diffusion of information and communication networks, see Nicolas Jeginer and Gerard Blanc's, *The World of Appropriate Technology: A Quantitative Analysis*, (Paris: Development Centre of the Organization for Economic Cooperation and Development, 1983).

[12] Mr. Kovall and Mr. Bhagat, Director and Assistant Director of CRS in Egypt, researched and designed this project together. Kovall's many years of experience in developing countries including Bangladesh and Morocco, just to name a few, and Bhagat's engineering genius and clear commitment to appropriate technology, were focused and realized in this highly successful project. For further information about this machine, including films, brochures and other related documents, please contact the Ford Foundation, CRS/New York, or CRS/Cairo, Egypt.

ROMUALDAS SVIEDRYS

A CONCEPTUAL FRAMEWORK FOR UNDERSTANDING TECHNOLOGY TRANSFER TO THE THIRD WORLD

INTRODUCTION

Every transfer of technology has the potential for a curious role reversal between donor and recipient. The latter can improve on the former and retransfer technology and commercial products back. The fact that such a "happy" outcome is infrequent only shows the complexity of the process and the considerable time required to achieve full mastery of most technology.

If a recipient is successful in transferring, absorbing, and mastering technology, the donor faces the possibility of losing jobs at home and markets abroad. This has happened in the textile industry of the developed world which today is reeling from foreign imports. In the European Economic Community, the industry is losing about 100,000 jobs a year due to imports. The South Koreans are beginning to flood the American market with VCRs, cars and steel-technologies that were successfully transferred from Japan. A number of newly industrialized countries have now become prominent as exporters of weapons. Some are selling military hardware only in the Third World and undercutting the expensive and complex products of the industrial nations. But others such as Brazil and Israel have emerged among the ten top sellers of weapons in the world and are even exporting their designs to the United Kingdom and the United States.

Common failures and such unique successes in technology transfer both call for a clearer conceptual understanding of the technology transfer process.

TYPES OF TECHNOLOGY TRANSFER

When a technology that was developed in an industrial society – a particular social, economic and cultural environment – is put to work in a different environment we get the process that is called technology transfer. This entails much more than just sending some machinery and

185

Edmund F. Byrne and Joseph C. Pitt (eds), Technological Transformation: Contextual and Conceptual Implications, 185–200.
© *1989 Kluwer Academic Publishers.*

starting it up. Indeed, the key to successful technology transfer is adapting the technology to its new environment. This modification to fit in better with the economic, social, physical and cultural environment often requires subtle changes in the recipient society, including the creation of local groups of experts able to initiate technical modifications.

Technical innovations that are required to adapt technology to its new environment are just the tip of the iceberg. Below these, we often see further transfers in the processes required to make the components, intermediaries and inputs, as more and more of the parts are manufactured locally. Process innovations also occur if the volume of production increases or if there is product innovation.

In the context of technology transfer, the meaning of "environment" is also much wider than is normally understood. By environment we mean much more than geography and climate. Geography and climate are themselves crucial in any agricultural technology transfers because crops are site and climate specific. But even here, environment includes additional factors such as social attitudes and beliefs, level of economic development, local, regional and national policies, availability of skilled human resources, business environment, social institutions and the many other factors that make up the cultural matrix with which technology must mesh and within which it must thrive.

The key to successful technology transfer has always been the local knack to adapt an imported technology to its local and new environment. The greater the similarity between the environment of a donor and recipient, the greater the potential for successful adaptation and thus successful technology transfer. For this reason, most effective transfers occur among industrial nations. And since the difference in environments is greatest between the Third World and the developed world, most failures occur when a technology which has been created and developed in an industrial society is transferred to the developing world. The more developed a country is, the better it is able to select, modify, and absorb foreign technology. Japan, Israel or Brazil are better able to negotiate skillfully for new technology with any donor. In case of failure to agree on terms, they have the capacity to reproduce or copy the technology in question. In these matters Third World countries are at a terrible disadvantage.

The definition of technology transfer as adaptation of technology to a new environment should remind us that there are three different types of technology transfer. In addition to (1) the transfer of technology from one society to another, there is also (2) the transfer of technology within a society, say from aerospace to medicine, or from one application to another, from one function to another. The rich technological history of the United States has many such technology transfers. A manufacturer of firearms transferred its process technology to the production of typewriters (Remington), and a bicycle manufacturer went into automobile production (Cadillac). Textile production led some companies into machine tools production. This type of technology transfer occurs very frequently, and yet it often requires considerable adaptation to the new environment. It is not an easy transfer; it requires ingenuity, creativity, talent, research and development.

Another type of technology transfer from one environment to another is (3) the transfer from research laboratory to commercial application. What works in the laboratory, or operates on a small scale must be transferred to society. Clearly, the financial, cultural and other differences between a laboratory and society at large are considerable. The technology must be made to be user friendly, reliable, safe, low maintenance, easy to repair, affordable; and such an adaptation is not easily or quickly done. A good example of this is the VCR which was developed for the American TV broadcasting industry to tape programs for delayed broadcast in different time zones. To put this product into our homes, the Japanese had to bring the unit costs down from $100,000 to $1,000, reduce the tape from two inches to half inch and still record good quality, miniaturize the machine, make it foolproof, and develop a large number of process technologies to gear up for mass production.

This is a fine example of technology transfer from laboratory to commercial success. Developed countries engage in all three types of technology transfer and it often is a matter of economics or national security whether something should be transferred from abroad or developed locally.

Types of technology transfer (2) and (3) are characteristic of technology-initiating countries. Thus, the long range goal of any technology transfer to the Third World is to develop small but dynamic local communities of research-minded engineers and scientists who can undertake these activities which are so much needed for the selection, adaptation, modification and eventual production of technology. While the latter is happening, the short range technological needs can often best be met by technical cooperation between Third World countries.

Such a procedure has an immediate economic advantage. It certainly is cheaper to get technology from other developing countries than from developed. Technology from such a source is already adapted to some of the conditions that prevail in developing societies: tropical climates, inexpensive labor, or lack of infrastructure. When Nigeria's railroads were near collapse due to mismanagement and maintenance problems, the Nigerian government decided to ask for help from India rather than Britain. An Indian group of experts ran a three year program on the Nigerian railway network that produced positive and long lasting results. And psychologically, it was easier for Nigerians to take criticism from Indians than from the British.

Technical cooperation among developing countries is reinforced by conferences organized and sponsored by various international bodies. The United Nations Development Programme (UNDP) sponsored one such conference in 1978 in Buenos Aires. It clearly showed to most participants from developing countries that technology transfer is tied to economic development which specifically requires the creating and nurturing of an internal innovative capability. It also showed that even developing countries can create sophisticated technology.

AN EXAMPLE: MEXICO DEVELOPS NEW STEEL-MAKING METHOD

All industrial societies use coke for the smelting of iron ore. A Third World country interested in the production of iron and steel would at first sight seem to have to transfer this technology from the developed world. Countries without coal, and specifically, without cokable coal, would be at a disadvantage because they would have to import coal or coke in order to work their steel industry. There is another option, namely to use a type of technology based on charcoal smelting. Indeed, one Brazilian steel manufacturer does this, but it results in rapid deforestation around the production site and increased costs in shipping charcoal from rapidly increasing distances.

Given that coal is not available in most Third World countries, their prospects for steel-making were dim. Mexico is a case in point. There are no appreciable coal deposits in Mexico. But Mexico has plenty of natural gas, and developed a process to reduce iron ore using natural gas instead of coke. The method is called direct reduction and consists of just two simple steps to convert iron ore into raw steel. The equipment used is simpler than the traditional method because the installation uses no coke, no pig iron, no blast furnace and can be built for less capital: from one half to two-thirds the cost of a conventional plant of the same capacity. And whereas there is a certain minimum size for conventional steelmaking plants, the direct reduction plant can be built on a small scale. Thus the direct reduction method is just what is needed in developing countries. It is a capital-saving technology which is better suited for small markets, and it can be distributed all over the country. This prevents overloading the meager transportation networks of developing countries first with raw materials and then with finished products. In addition to requiring less capital, production capacity can be added in small incremental steps, as demand increases. By contrast, the coke steel plants require huge size to realize economies of scale.

This technology is a major attraction for all developing countries. It is true that not all have natural gas, but it certainly brightens prospects for all OPEC and oil producing countries. Natural gas is often flared off as waste-product and it now becomes a valuable raw material that can be used to produce steel directly where the natural gas is available.

The first Mexican plant was built in 1957. The process has been exported to other cou..tries, including Venezuela, Iraq, and Indonesia. It is attractive even to developed countries that have plenty of conventional steel making capacity and no natural gas. Converting to direct reduction can increase a conventional blast furnace capacity by 30% because more iron ore can be put in, given that coke and limestone are no longer needed. Natural gas can be imported, which makes the potential for retrofitting truly staggering since 97% of the world's steel making capacity is based on coke reduction blast furnaces.

The Mexican direct reduction case shows that developing countries can come up with solutions that are more appropriate to their environments than technology used in the developed countries. Adopting the steel technology of developed countries, the large scale production technology, may on occasion still be a good idea even if the production capacity exceeds local demand for steel. Excess production may be exported, if other developing countries similarly built large scale automobile plants, home appliances and other steel consuming industries on a larger scale than a purely national market. But such an approach would also require political and economic cooperation among developing countries. The latter is more complicated and difficult than technical cooperation.

THE TRANSFER PROCESS: A MACRO ANALYSIS

How does technology transfer work, and when do we know that it is successful? What determines how rapidly technology transfer takes place? It is no doubt discouraging for developing societies to discover that the process can take about a century from beginning to full mastery. It is a learning experience with distinct stages and changing criteria of success.

The transfer of firearms technology from Europe to North America began while it was still a British colony and took about one hundred and fifty years until designs and production methods developed in America migrated back to Britain. Today, if everything is done under great pressure (say, national security pressures), the time required may be cut to one half or one third. Japan's transfer of naval technology from Britain took about 75 years to surpass Britain in terms of excellence and numbers of battleships and aircraft carriers built. Israel's aerospace industry has taken about 50 years to develop to maturity.

The reason technology transfer takes so much time is that it presupposes social and economic modernization. In Japan, naval technology required the development of a steel industry, together with modern explosives, shipyards, as well as the establishment of engineering education such as naval engineering. In Israel, aerospace meant also development of avionics, computers, wind-tunnels, and post-graduate education in a large number of technical subjects. Even so, more than half of the Israeli designed Lavi fighter plane will be built by U.S. subcontractors.

The Japanese and Israeli experiences provide a time frame for technology transfer and can serve as a model for today's developing countries – provided they do everything right and work hard at it. It must be understood that technology transfer is a learning process which needs periodic rewards to keep the student eager and motivated, and like any learning process, it can have setbacks and failures even with superb human resources available and involved. Even small initial successes are crucial for the continuation of the transfer process, while initial failures dampen motivation to continue the process and make future learning more difficult.

The transfer of firearms technology to the United States had one of its early successes in the production of spare parts (replacing broken stocks with locally made ones). Later, the ability to copy the hunting rifle from

Central Europe was a tremendous confidence booster. Improvements in the hunting rifle led to the local design known as the Pennsylvania-Kentucky rifle. During the nineteenth century, American gunsmiths pioneered new manufacturing techniques. The idea of interchangeable parts was transferred to other mass-produced goods. Experience gained in firearms production was transferred to the making of other products that required similar processes. And by the time of the 1851 London Exhibition, the American system of production was being admired in England. Samuel Colt set up a factory in Britain to manufacture his weapons on American-designed machines for which he had to import American operators.

There are distinct stages in the technology transfer process. Their number is a function of how much detail we are interested in looking at. As a first approximation, a three step sequence is sufficient to track a recipient's progress toward full mastery of transferred technology, or what also has been called a *technology-initiating* society. The three stages are transfer of artifacts, transfer of designs, and transfer of capacity to make local designs.

Transfer of artifacts. For a developing country the process begins with the simple transfer of machines, materials and techniques associated with those machines and materials. The recipient is a passive consumer of technology produced elsewhere by others. Sometimes the imported technology is used differently, or its impact on society may be different, or it may require the transfer of new patterns of behavior, such as maintenance. Minor adaptation is usually required, as well as careful choice from a range of available technologies, suppliers, and options. Too many Third World countries remain at this stage of technology transfer and are unable to see beyond it. Too many are too small to have their own computer or automotive industry. Too many lack the resources it takes to develop a new generation computer or sophisticated telecommunications hardware. Even some developed countries remain at the artifact stage by choice. Switzerland, for example, considers itself too small to have its own automotive industry.

Transfer of designs. A country that has reached this stage in an area of technology no longer needs to physically transfer (buy, import, smuggle) artifacts. It has the means to copy foreign technology, and all it needs to transfer are the design drawings or the production formulas.

When a production license is not available, they can buy an artifact and copy it. This has been called reverse engineering, although pirating is a better term. Industrial espionage is a viable and inexpensive method of transferring technology only if there is ability to copy. In any case, it is a sign of rapidly maturing technical capability, and an obvious sign of progress, no matter how alarming to the donor of technology. The recipient who has reached this stage still remains dependent upon technological knowledge and designs produced elsewhere, but it is rapidly approaching a capacity to modify and improve on foreign designs. In many areas of military technology the Soviet Union is currently operating at this stage of technology transfer.

Transfer of the capacity to make own designs. This marks the approach to full mastery of technology, a technology-initiating society. Among other things, it requires the transfer or development of a science and technology post-graduate educational system, together with a vigorous research and scientific tradition. It requires many concurrent developments in related technologies, as well as general economic, social and political modernization. It requires increased social mobility to attract talent into these areas of science and engineering. Whereas in the design transfer stage design plans and blueprints are all that is needed to produce technology locally, at the capacity transfer stage just reading scientific papers published abroad, attending scientific meetings, or visiting foreign laboratories is sufficient to start local activities that lead to production of knowledge and locally designed technology. Western Europe, Japan, Israel, and the United States have scientific relations that illustrate this stage of technology transfer with numerous examples. Industrial espionage and the hiring of skilled people in the donor country to serve as "antennae" are also common occurrences.

No matter what the degree of magnification, the various stages overlap a little at the edges, blending into one another. On the basis of this macro analysis, however, it is helpful to proceed to describe in greater detail the progress along the road to full mastery of technology.

STAGES IN TECHNOLOGY TRANSFER: A TEN-STAGE MODEL

(1) *Import artifacts*. Learning how to use artifacts is usually no problems in most societies. When economic resources are plentiful, there is no

need to choose carefully or agonize over learning maintenance since these things can be left for foreigners to do. Saudi Arabia during the oil boom years is an example which comes to mind. When money is plentiful, there is no further need to work toward technology transfer.

Yet all countries must make choices; some buy new, others second hand industrial products and equipment. Even in poorer countries it has often been true that technology transfer at this stage has been associated with a large influx of foreigners whose skills were needed to run the complex artifacts that were imported – modern combat planes, nuclear power plants, petrochemical complexes and so on. Learning how to use imported artifacts well can constitute that initial small triumph that a learning process requires.

(2) *Service and maintenance*. This is the next hurdle that any recipient must successfully negotiate. Maintenance may be a very difficult concept to transfer; it often leads to a serious clash with established cultural norms of the recipient. Many transfer projects die a natural death at this stage due to additional factors such as lack of hard currency to buy spare parts, or a deliberate embargo. But primarily, Third World countries are littered with the countless rusting skeletons of machines for which spare parts were included neither in the initial nor in the operating budgets. Learning to maintain artifacts and being able to do so properly is an enormous achievement.

(3) *Local assembly using imported parts*. This step is often visible in automobiles, airplanes, tractors and electronic goods. Very often these goods are assembled in special economic zones and exported back to the developed countries. The northern border of Mexico is a prime example of this approach. Slight modifications are introduced if assembly is for local markets. Offshore assembly using low labor costs constitute a source of exports that generates hard currency to finance further technology transfers, purchase of spare parts, and pay for foreign experts whose skills are needed.

(4) *Spare parts production*. Spare parts production triggers large additional technology transfers in process technologies which often must start again at stage (1) and go through maintenance learning, and so on. Local entrepreneurs begin to appear to exploit opportunities. Spare parts production may entail local production of chemicals that are

needed, foundry work and so on. The level of technology mastery is very different when a country is producing about 10% of the component parts of a car than when it is able to produce 50% or 90%. It takes time to move from a few spare parts produced locally to a situation where all parts can be produced locally. At this stage, import substitution can be a powerful economic incentive to continue the technology transfer process.

(5) *License for local production.* Often a subsidiary of a multinational company will produce locally, or grant a license for a local company to produce their product. If no license is granted, then local companies can copy and produce for local markets or for export. Copying of foreign designs should be considered as a sign of progress in the process of technology transfer. It is a clear signal that the mastery of technology has begun. It is an indication of increasing confidence, and it creates the ability to substitute for imports. For donor countries this is the first warning flag of competition ahead. Licensed or unlicensed production often incorporates minor improvements and modifications on foreign designs. Korea, Brazil, Taiwan, Hong Kong, Spain and a large number of countries can increasingly produce under a license. Israel, in response to the French embargo of 1967, chose to produce the Mirage fighter bomber without French permission.

(6) *Improvements on foreign designs.* If improvements are universal improvements rather than just reflect adaptation to local conditions, advances towards local designs with export potential will be possible. This is clearly a further realization of the potentials present in stage (5). Together stages (5) and (6) constitute the pivotal stages in moving from passive reception to active mastery. Following the French embargo, the Israelis switched to American jet engines and produced a copy of the Mirage locally. Gradually, they added hundreds of modifications that resulted in the *Kfir*, which was exported to several countries. One squadron is currently used by the U.S. Navy as an "aggressor" squadron to provide realistic training for Navy fliers.

(7) *Major improvements and modifications of foreign designs.* With this stage, the ability to transfer technology from one application to another, becomes increasingly necessary. A growing pool of locally educated technicians and engineers must be created. The latter entails graduate

level educational institutions, established a decade or two earlier, to allow them to begin gearing up with local research done by locally trained people. The ability to modify process technologies when product innovations are made is also required. Such developments were well illustrated in the Israeli experience. The second generation Kfir had a canard developed locally, improved avionics and a host of modifications that incorporated lessons learned in the 1973 war. The Department of Aeronautical Engineering had been established in 1953, and its wind tunnel was used for some of the testing and research.

(8) *Capacity to make one's own designs*. Now local research and developments centers are able to work on original research projects. Local designs made by locally trained people are the long term goal. The first designers often get their training abroad, but then return to teach at the graduate level. The designer of the canard for the Kfir had received his Master and Ph.D. (1958) at Polytechnic University in Brooklyn.

(9) *Self-sustaining growth of innovations and technology-initiating status*. Clear signs of an incipient technological autonomy begin to manifest themselves. There is the creation of forward and backward linkages in both economy and technology. Locally produced technology begins to sell abroad; the country begins to get patent and royalty payments on its technologies, and to complain about other upstarts who copy its designs illegally.

(10) *Capacity to compete internationally*. At some point a country becomes an acknowledged source of technology which constitutes a "graduation diploma" of recognized technological excellence. Germany and the U.S. were such at the turn of the century; Japan clearly entered this stage about ten to twenty years ago (depending on the technology), and Israel is about to get this recognition in arid zone agricultural technology, water resource management, solar energy, biotechnology, electronics, aerospace, and data processing.

ANALYSIS OF THE TEN-STAGE MODEL

In the above model, the increased number of stages still produces overlap between one stage and another. At what point the Israelis moved from stage (5) to (6) is a debatable question. In 1959, their test pilot Danny Shapira flew the French Mirage III prototype, and he and others made many useful suggestions that were incorporated into French production line. Then, due to the very intensive use the Mirage III planes received in the Israeli Air Force, where they logged far more flying hours than Mirages in the French Air Force did, most of the teething problems were discovered by the Israelis. The resultant design changes were incorporated into later production back in France and even sold to Arab countries. When the French embargo cut the Israelis off from spare parts and further purchases of Mirage planes such as the improved Mirage V based on Israeli design specifications, the Israelis opted for local production without a license. This copy of the Mirage, called the Nesher, flew in combat in 1973, and was later used by Argentineans against the British in the Falkland Islands in 1982. But while putting this plane into production, the Israelis simultaneously started to design an improved Mirage/Nesher. Should the resultant plane, the Kfir, be considered a local design or was it a case of major improvements to a foreign design? Crossing the boundary from one stage to another is seldom clear cut. But, the value of this schema is to offer major landmarks in the transfer process and to hint at the major reasons why it takes so long to achieve technological mastery. It also points out that the slowness of the process cannot be attributed primarily to some (real or hypothetical) unwillingness of a donor to share technology with the recipient.

These ten stages could be expanded into fifteen or any desirable number that provides the amount of detail desired in a particular area of technology. But in relation to the initial macro analysis, it is clear that the transfer of artifacts is covered by stages (1), (2), and (3). Stage (4) is astride transfer of artifacts and transfer of design, because spare parts are often copied with or without a license. Transfer of designs is covered by stages (4), (5), (6) and (7), while transfer of the capacity to create one's own designs is covered by the remaining stages (8), (9), and (10).

Sometime during the transfer of design stage, it becomes increasingly possible for a country to practice the second type of technology transfer – that is, to transfer technology from one application to another, from one function to another. This type of transfer is also called horizontal technology transfer.

The third type of technology transfer – that is the transfer from idea to commercial product – entails production of new technology, and such a feat must be fully developed during the stage that we called *transfer of capacity to make one's own designs*. This is the stage during which the transplanting of science and its institutions must be accomplished, and a vigorous research tradition established. Basic research produces new knowledge which later becomes technology as applications are found and the technology developed. The capacity to create modern technology, or technology that is appropriate to local needs, constitutes a significant difference between developed and developing countries. By the time a developing country reaches this stage of technology mastery, no doubt, it will already be a developed country if it has been able to accomplish the same development in many areas of technology.

CONCLUSION

The successful transfer of technology requires various degrees of modernization in the recipient society. At the same time, technology transfer is no panacea to the problems of economic development. Technology transfer is getting easier to the extent that today there are many competing donors of technology. But it remains difficult and successful for only a few recipients because of the non-technological parameters which play such an important role in both economic development and technology transfer. Only when technology is *the only missing ingredient* in the development process, can technology transfer help stimulate economic growth and development. Otherwise, technology transfer can be, as it so often is, disappointing in its results.

The transfer of technology is intimately connected with economic and social development, and quite often policy failures can frustrate all of them. If a government overvalues its local currency, local firms will have an incentive to import foreign equipment rather than develop local technology. This can condemn a country to remain at the artifact transfer stage rather than create gentle pressures that move a society to the next and succeeding stages.

On the next level of technological maturity a policy that works well for the protection of local producers may keep them protected far too long and prevent their emergence as exporters to international markets and developers of improved designs. Substitution of imports requires one set of policies (tariffs, subsidies, etc.) that is in conflict with policies designed to help export locally designed technological products, and for successful competition in international markets.

To reach the stage of the capacity to make its own designs, a country requires the transplanting of a graduate level education system and a vigorous scientific research tradition. Since such a process takes considerable time, the first steps must be taken long before a country reaches technological maturity. Science, like technology, is a social product and process that is vitally affected by its institutional settings. And one of the greatest problems of Third World countries is that as they start to develop in science, their scientists migrate to developed countries. Developing countries import research (embodied in technology or in designs) while exporting scientists and engineers.

Why do scientists leave Third World countries? They leave because their society does not provide adequate conditions for the growth of science and technology. This is truly a Catch 22 situation and a problem which remains on the agenda of all developing countries – how to transform an environment that is alien to science into one that is attractive for the work of local scientists.

How can developing countries produce first-class, world-class science? And how does one create the missing research and science tradition? Although there are programs for the return of talent to various developing countries, so far, results have been quite meager. In the current century, only two countries have managed to join the club of scientific heavy-weights. They are Japan and Israel. They now have the distinction of being among the 20 largest science producers. Israel has achieved this despite a population of approximately three million. That is a per capita rate that is approximately one hundred times larger than is produced in the developing countries.

The process of transplanting science to Israel took almost one century. It was made possible by the existence of a global network of Jewish scientific talent that emigrated to Israel and settled there, and to wise government science policies that favored travel and contacts with the outside scientific communities, as well as provided the resources for scientists to work with. For most developing countries the transplanting of science is the greatest long-term obstacle to overcome. How many will be successful no one can be sure, because the precepts and mores of science are transmitted like any other social tradition. It is not clear whether the modern scientific tradition can be disassociated from Western civilization nor whether once it is imported the other necessarily follows. Nor is it at all clear that the Marxist model of economic development which is so much sought after by leaders of the Third World will be able to provide the social milieu to develop a vigorous set of research and scientific traditions in the cultures of the Third World.

BIBLIOGRAPHIC NOTE

My interest in technology transfer was first aroused by personal experience, during a summer stint with a French company, Tissot & Cie., which was involved with several other companies in building the first steel mill in Colombia. Five years of teaching in an engineering school in Colombia reinforced this interest.

Melvin Kranzberg's paper read at the 1970 Pont-A-Mousson colloquium and published in *L'Acquisition des techniques par les pays non-initiateurs* (Paris: Editions du Centre National de la Recherche Scientifique, 1973) was influential on my own thinking about technology transfer.

David J. Jeremy, *Transatlantic Industrial Revolution: The Diffusion of Textile Technologies between Britain and America, 1790–1830s* (Cambridge: MIT Press, 1981) is basic to the development of technology transfer model discussed above, as well as Stephen Howart, *The Fighting Ships of the Rising Sun. The Drama of the Imperial Japanese Navy, 1895–1945* (New York: Atheneum, 1983) which provided a confirmation of the model in the field of military technology. Two field trips to Israel, as well as bibliographic research, provided data for my case study of aerospace technology transfer to Israel which is in a final stage of preparation.

I have always found Nathan Rosenberg's work excellent, and will cite two of his book as most relevant to technology transfer. His *Perspective on Technology* (Cambridge: Cambridge University Press, 1976) has been used as a textbook in my technology transfer course. His more recent book, *Inside the Black Box: Technology and Economics* (Cambridge: Cambridge University Press, 1982) brings together papers that were published in various journals.

Nathan Rosenberg was one of the editors of *International Technology Transfer. Concepts, Measures, and Comparisons* (New York: Praeger, 1985) which consists of eight papers presented at a conference held in June 1983. The wide geographical dispersion and the great variety of technology transfers considered can be complemented with the excellent study of just two technologies, textiles and pulp industries, carried out by Michel A. Amsalem, *Technology Choice in Developing Countries* (Cambridge: MIT Press, 1983).

LAN XUE

APPROPRIATE TECHNOLOGY IN TECHNOLOGY TRANSFER:
A VIEW FROM THE PEOPLE'S REPUBLIC OF CHINA

1. INTRODUCTION

As early as 1938 J.D. Bernal asserted in the preface of *The Social Function of Science* that:

Any discussion of the application of science necessarily involves questions of economics, and we are driven to enquire how far the various economic systems now existing or proposed can give the opportunity for the maximum application of science for human welfare.[1]

The assertion is still true after almost 50 years, except that we can substitute "technology" for "application of science" and reverse the first half sentence to read "any discussion of economics necessarily involves questions of technology".

In fact so many have taken for granted the key role of technology progress in economic development that few have ever thought of doing quantitative research to support this assumption. Fortunately some research has been done and here let us survey briefly these efforts to quantify the importance of technology in the economy.

Schmookler suggested that the growth of the national product in the U.S.A. in the 70 years ending in 1938 was due in equal parts to the growth of resource input and to the growth of efficiency in using them.[2] Abramovitz and Kuznets[3] supported this view which was later on strengthened by studies of Solow and Kendrick. According to them, from about 1900 to 1920 technical progress or productivity advance contributed 1 percent a year to the rise of U.S. national output. But between 1920 and 1950 this contribution rose to 2 percent a year. According to Hogan[4] some 90 percent of the growth of productivity in the non-farming sector of the U.S. was due to technical change.

Though some people might question the accuracy of these figures or even the approaches by which these figures were obtained, there should be no doubt about the general conclusion of these studies – advance in

201

Edmund F. Byrne and Joseph C. Pitt (eds.), Technological Transformation: Contextual and Conceptual Implications, 201–225.
© *1989 Kluwer Academic Publishers.*

scientific knowledge and progress in its technological implementation have added very appreciably to economic performance and to the raising of living standards all over the world. Never before in history has technology played such a dominant role in economic development, promising so much good and threatening so much evil as it does today. It seems without exception that economically developed countries are advanced in technology and economically underdeveloped countries are comparatively backward in technology.

A study conducted by the Organization for Economic Cooperation and Development (OECD) in 1970 identified 112 significant innovations in the twentieth century; all emanated from developed countries (DCs), with the United States contributing 60 percent; the United Kingdom, 14 percent; and West Germany, 11 percent. The DCs are estimated to account for 97 percent of world expenditure on R&D, although this must be equalized to allow for the lower salaries paid scientific manpower in less developed countries (LDCs); LDCs account for 13 percent of world scientists and engineers engaged in R&D.[5] Of the three and a half million patents issued in 1972, only six percent were issued by LDCs.[6]

Based upon the above two facts, namely the importance of technology in economic development and the domination of technology innovations by DCs, it is no wonder why technology transfer from DCs to LDCs has been considered an important way for international development. Past experience, however, has shown that technology transfer between DCs and LDCs is far less successful than people had expected. The reasons for this failure have long been the topic of debates. The notion of "Appropriate Technology" is, in part, the result of these debates.

In this discussion of appropriate technology (A.T.) in technology transfer, I will first try to clarify some confusions about the term appropriate technology and give some criteria of technological appropriateness. Then I will analyze the reasons why sometimes technologies transferred to LDCs are not appropriate, and finally some suggestions will be made on how to improve the situation.

2. APPROPRIATE TECHNOLOGY

2.1 Review of Semantic Discussion on A.T.

Professor Max Dresden of the Institute of Theoretical Physics at SUNY, Stony Brook, asked me "What is appropriate technology?" when I met him this summer and told him that I was doing a research project on A.T. While I was trying to explain what A.T. was, he became impatient. He said "Maybe we can spend some time later especially discussing this subject. But I doubt whether we can find out what A.T. really is." Indeed maybe we can never find a clear, right-to-the-point answer which is what he wanted to this question, because there are so many different words and terms that we have heard over the years. Some call it "gentle technology", "alternative technology". Others call it "low-cost technology", "intermediate technology". The exact meaning of and differences among these terms are the subjects of lively, if somewhat inconclusive, theoretical debates. Here let's briefly review some of these terms.

Basically there are two groups of opinions. The first one is mainly concerned with the problems LDCs face when they transfer technologies from DCs or develop technologies themselves. The terms "low-cost technology" and "intermediate technology" might be categorized in this group. The ideas under these terms have been vividly expressed by Dr. E.F. Schumacher in his influential book *Small is Beautiful*, which, perhaps more than any other, has contributed to spreading the concept of intermediate technology, both in the LDCs and DCs.

To illustrate what are intermediate technology and appropriate technology Nicolas Jequier gave us some examples.[7] An example of intermediate technology is the gari machine developed in Nigeria. Gari, a dehydrated cassava product, is a staple food in most of West Africa. The traditional manual preparation method for extracting the prussic acid from the cassava root was brought to West Africa in the 18th century by former slaves from Brazil who had borrowed it from the local Indian population. With the rapid rate of urbanization, industrial production methods were required, and a modern large-scale technology was developed in the 1960s by a British firm in Gambia in co-operation with Nigerian technologists. Parallel to this, a smaller scale and somewhat simpler technology was developed at the time of the Nigerian Civil War on the Biafran side. The capacity of this intermediate plant was smaller, but the total investment required for the same output was at least four times lower and its profitability substantially higher than its modern counterpart.

The solar pump developed by a French firm, in co-operation with the University of Dakar, is probably a very good example of A.T. It uses a widely available source of energy – the sun – to provide villagers with a scarce but vitally important commodity – water. Although it is technically quite sophisticated, it blends rather well into the social environment. It requires virtually no maintenance and seems to have a significantly long working life. In the same way, a number of the technologies developed or popularised by the Brace Research Institute in Canada can be considered as particularly appropriate. Given these examples, Jequier concludes that:

What these examples suggest is that at this stage, the delineation between these various concepts is still in a state of flux. Appropriate technology is very close to, but not entirely identical with intermediate technology, and a low-cost technology, while often particularly appropriate to the condition in a developing society, does not necessarily always meet the criterion of appropriateness. In fact, each of these concepts might be viewed as a set of overlapping but nevertheless distinct areas, the frontiers of which are rapidly changing under the impact of recent experiments, new innovations and progressive changes in perspective.[8]

While the first group of opinions addresses technologies used, developed, or imported by LDCs, the other group of opinions, mainly originated from DCs, emphasizes the need for much greater attention to the ecological impact of new technology and to real needs of society.

This group of opinions might be represented by that of Mr. Sim Van Der Ryn, Director of the California Office of Appropriate Technology. He sees three characteristics of any A.T., namely syntropy, coherence, and diversity.[9]

Syntropy, according to Mr. Van Der Ryn, is the tendency to minimize the amount of waste heat in converting from one form of energy to another or in performing useful work. All forms of energy conversion, be they natural or man-made, are entropic. Natural ecosystem processes are inherently syntropic and efficient, that is, they capture solar energy and convert it to other usable forms of energy through complex chains of biophysical events. A.T. matches the properties of energy sources with end uses to achieve desired results with the least waste.

A second criterion is coherence. What this means is that a particular technology is in tune with and can coexist with the syntropic behavior of natural systems. Only in this way can a high quality of life be sustained. A.T. seeks a higher, more mature and scientific integration of human needs with ecological necessity in which there is no waste, since waste is only a resource out of place, and each byproduct feeds into another system.

The third criterion for an A.T. is diversity. What diversity gives us is the possibility to adapt to localized and changing conditions. We have seen that ruthless simplification and uniformity in a technology imposed for short-run advantage, an organizational convenience, has disastrous effects as conditions change. Again the principles of diversity follow from an observation of efficient natural process. A diverse, decentralized technology is adaptable to change and more likely to be syntropic.

In general, this group of opinions tends to see A.T. as small rather than big; as elegantly simple as opposed to dumbly complex; as crudely right as opposed to precisely wrong; decentralized rather than centralized; low cost rather than expensive; and employing capital and fossil fuel sparingly. Also, Mr. Van Der Ryn argues, A.T. does not mean low technology, as opposed to high technology. Much modern technology, such as that used in communications and electronics, lends itself to small-scale applications, and enhances the prospects for diversification and decentralization. However, communication and electronics are certainly not examples of low technology. The implementation of A.T. is a way of providing non-inflationary growth in a diversified responsive society. Using A.T. provides for sustained yield from our natural resources and gives people greater individual control and responsibility, a buffer against disruptive and uncontrollable results of fossil fuel shortages, and the skyrocketing costs of immature, highly entropic technology.

2.2 Criteria for Technological Appropriateness

As the discussion on A.T. develops, more and more people realize that a particular technology, either intermediate technology or gentle technology, is not end itself but means for achieving economic, political and social goals. A.T. should not be any particular technology; it represents a way of thinking or a strategy in using technologies to reach those goals. This realization has led people to the discussion of the criterion of technological appropriateness. Here, based on the discussion by Richard S.

Eckaus,[10] let us take a look at some of the most often cited criteria.

The first criterion is maximization of net national output and income. It is perhaps the most commonly applied either implicitly or explicitly as a primary goal of development because, in general, it provides the largest sum of money and goods to meet various needs in a country. The rules for technological choice to maximize the national output and income at any time require physical and economic efficiency in each productive establishment and in the economy as a whole. These rules are embodied in the methods of cost-benefit analysis that can be applied to new projects and technical decisions and in the normative overall planning techniques that have been developed. A significant limitation of this criterion, however, is that it avoids the essential questions of what should be the composition and distribution of national output. Nobody can be sure whether the national output is used directly or indirectly for the right social purpose and distributed according to the right social rule.

The second criterion is reduction of unemployment. It has generated the greatest concern about the appropriateness of technological choices. So there is no wonder that employment generation is considered one of the key criteria of appropriateness of technology. Here two aspects are mainly concerned. The first one is the number of jobs created, which is of the utmost concern for LDCs. Despite the substantial amount of investment that has occurred in most of the developing countries, their unemployment problem has not been relieved, partly because their labor force growth rate is high compared with their growth rate of unemployment. The second aspect is the nature of the work and the quality of employment. In addition to key social and human considerations, there is a growing desirability of paying attention to the scientific and technical content of jobs. The higher this content, the greater are the prospects for indigenous and autonomous technological growth. So this concern is very important even though under current conditions, LDCs might not yet have the political and economic flexibility to make choices on this basis.

The third criterion is regional development. It is simply economic, political, and other social goals specified for a particular locality. If there is any difference between such regionally specified goals and those set forth at the national level, it is that the regional goals may be described in terms relative to the average of the nation as a whole or some particular 'advanced' region. Differential rates of regional development are responses to differential availability and use of resources. It is highly valued, therefore, to maximize local integration and resource use by making the right decision on investment in particular types of production. The greater the number of local interconnections and linkages a technological process has, the greater the local value added to the product, and the larger the local benefit.

Another important criterion is potential for technology adaptation. The greater such a potential, the more attractive the technology. Adherence to the pure engineering, efficiency standards had demanded that the receiving environment do all the adapting in order to properly accommodate the incoming high technology. The result of such efforts led to either failure or islands of isolated high-technology production within regions of much lesser technological sophistication. While this requirement of potential of technology for adaptation is important and necessary, overemphasising it might be harmful to the development of the whole society because modern technology is not only a means to achieve economic progress, but also a driving force for social changes. The adaptation of a new technology, even in DCs, often necessitates change of the society to accommodate it. This change, along with others, constitutes what we call the development of society. So the process of technology adaptation needs the interaction between the technology and the local conditions within which the technology is adapted.

Maintaining the quality of the environment is another important criterion. The industrial pollution and deterioration of our ecological system accompanying industrialization have long been a concern both in DCs and LDCs. While in DCs the use of hazardous chemicals in industrial processes has been strictly banned, LDCs have viewed the need to increase food and textile production as more important than avoiding acid rains or DDT's harmful effects and therefore, they do not give high priority to environmental issues. The consequences of this negligence have been revealed in several major industrial accidents – including the explosion of a liquefied gas tank in Mexico that killed 500 people, and the leak of deadly methyl isocyanate from a Union Carbide pesticide plant in Bhopal, India, that killed more than 2500 people. These accidents aroused concerns about the introduction of hazardous industrial processes in LDCs, which often lack the infrastructure needed, including a reliable supply of electricity and skilled labor, to run such facilities safely. So the ability to improve the quality of the environment should be another criterion of technological appropriateness. In fact, except for technologies especially created to control pollution, most technologies have not been so powerful and perfect as "pollution-free". But at least they should meet the reasonable standards set up by international organizations.

Though we could add other criteria (to cite just a few examples, maximization of availability of consumption goods; balance of payments relief; the property of diffusion; and so on), we have already covered some major ones in our discussion. What should next be noted is that these criteria should be used in a comparative context rather than an absolute context. It makes no sense to use these criteria to justify a technology itself and come to the conclusion about whether it is appropriate or not. Only by comparing it with its alternatives, can we say whether a technology is more or less appropriate than others. In doing so two major difficulties often emerge. The first one is that these criteria are alternative or competitive rather than complementary. For example increasing employment and reducing unemployment may or may not be compatible with either output and income maximization. Employment creation in the short run may well compete with employment creation over a longer period if the former results in the sacrifice of investment that would create employment opportunities in the future.

Partly as the result of the first difficulty, the second one is encountered when a technology is justified as appropriate by certain criteria and inappropriate by some other criteria, while for its alternative, it might be just the opposite. One obvious example is automation in production. It is appropriate by the criterion of maximizing output or profit but inappropriate by the criterion of reducing unemployment. The old unautomated technology would be the opposite: appropriate by the criterion of reduction of unemployment but inappropriate by the criterion of maximization of output. The underlying conflicts are the confrontations of different values, which can not be escaped and for which there is no analytical resolution. It is in this sense that our discussion about A.T. is so meaningful.

3. WHY INAPPROPRIATE? – REASONS FOR THE FAILURE OF
TECHNOLOGY TRANSFER FROM DCS TO LDCS

If technologies were like different tools in a tool box and could be taken easily as needed, then our discussion could have stopped at the last section. Unfortunately, technologies are more like the basic blocks, which can be used to build different structures. These structures could be very stable and beautiful, they could also be ugly and unstable or they could even collapse. Our discussion in the following part will explore the reasons why some structures LDCs built aren't so successful. In other words, why are they inappropriate?

3.1 Lack of a Long-Term, Consistent Strategy and the Corresponding Policy in Technology Development

Although we can attribute the failure of technology development in LDCs to many factors, lack of an appropriate long-term, consistent strategy and the corresponding policy is probably the fatal one.

A very good example of how government policy can affect the development of technology can be provided by the comparison of Japan's practice and China's experience.[11]

In the late 1940s and early 1950s, Japanese industrial policy was devoted to putting into place basic industries – cement, chemical fertilizer, electricity and steel. The economy was minuscule, protected as much by the indifference of others as it was by its own highly restrictive regulations. Any inputs of technology to Japan in those years required specific government approval. The approval procedure was onerous, detailed and quite directive. The principal effect of the Japanese government's early controls was first of all to avoid monopoly – that is, to prevent any one Japanese company from having a single technology. This was achieved through issuing licenses in such fashion that if one firm was getting technology another firm at the same time or soon after was allowed a competing technology from another source.

At the same time the Japanese government was clearly playing an important role in the balance between seller and buyer. Japanese companies in the period of the 1950s were desperate for technology and were prepared to pay almost any price. The Japanese government intervened directly in the negotiations by putting a ceiling on royalty payments, by working to limit the duration of the arrangements, by attempting to limit the scope of the agreements, and in particular by trying to reduce the export restrictions that are often embodied in technological agreements of this sort.

As Japanese buyers became more powerful, more nearly balanced with the sellers, and as the foreign sellers of technology were anxious to sell, the Japanese government's restrictions gradually relaxed. In the late 1950s and the 1960s the approval system stayed in place and was still thorough, but there was a much broader application in the sense that licenses of consumer product technology and licenses on duplicate technology received rapid approvals. By the late 1960s the approval apparatus was discontinued and by the early 1970s they were essentially completely removed except in a very few sectors.

The success of Japan's economy proved the success of its industrial policy. In comparison, let's take a look at China's experience in introducing foreign technology since 1949 – an experience commonly described in terms of four waves.

The general pattern of the four waves can be seen in Table 1., which shows China's annual imports of machinery and equipment from 1952 to 1982.

Table 1.[12]   China's import of machinery and equipment (US$ million)

Year	Imports	Year	Imports
1952	193	1967	335
1953	276	1968	235
1954	381	1969	214
1955	411	1970	398
1956	545	1971	481
1957	566	1972	524
1958	715	1973	797
1959	933	1974	1605
1960	840	1975	2013
1961	272	1976	1716
1962	102	1977	1171
1963	100	1978	2033
1964	162	1979	3832
1965	302	1980	5352
1966	443	1981	4661

As Table 1. shows, the first wave of technology transfer occurred in the 1950s, when 256 complete plant projects from the Soviet Union and a similar number of turnkey projects from Eastern European sources provided the technological core of China's First Five-Year Plan. One Western analyst has called this "the most comprehensive technology transfer in modern industrial history". But it came as a great shock when, in 1960, the Soviet Union "tore up contracts, withdrew its experts, and discontinued the supply of equipment". Fortunately for China, most of the Russian projects were already completed at that point. Furthermore, the East European countries only partially followed the Russian lead, which also cushioned the blow to some extent.

With Russian supplies of machinery and equipment cut off, China turned to alternative sources. In late 1962 China's second wave of technology import started from small beginnings. This wave grew rapidly for about three years during which China purchased a variety of complete industrial plants from Japan and Western Europe. But this growth of second wave was severely disrupted by the Cultural Revolution.

The third wave of plant and machinery imports, developed in the early 1970s. The China National Technical Corporation was revived in 1972 and in 1973 the State Council approved a plan to spend US $4.3 billion on imported equipment over a four-year period, known as the "Four Three Programme (the term came from the planned level of expenditure)". Among the various commentaries on this "Four Three Programme" an article by Chen Huiqin in 1981, is one of the most detailed and astringent criticisms.[12] Chen's criticisms were focused on three problem areas:

a) Failure to meet construction schedules;
b) Operating at less than full capacity;
c) Poor return on investment.

He cited the 1.7-meter steel rolling mill at Wuhan as an example of his criticism:

The 1.7-meter steel rolling mill at Wuhan, which required about 16 percent of the total investment (by implication, about US $560 million of foreign exchange, plus any complementary investment in RMB), was adversely affected after it was completed in December 1978 because the supply of electricity was inadequate. Only after the two provinces of Hubei and Henan integrated their power grids in May 1979 was it possible to alleviate this shortage of electricity. However, at the present time the utilization rate at the rolling mill has still only been raised to about 20 or 30 percent, because of endemic internal problems: inadequate supplies of ore and pig iron, deficiencies in smelting operations, delays in completing the processing facility for silicon steel, heavy reliance on imports for spare parts, etc. Even in a good year profits have been only a few tens of millions of RMB".[14]

In spite of the problems and shortcomings in China's third wave of technology transfer, another big wave of technology imports was launched. Within the single year of 1978, the worth of foreign contracts signed was twice as much as the total sum of the five years before 1978.[15] In the eyes of the critics, the unprecedented surge of technology imports was a classic case of "more haste, less speed". One thing in common in the four waves is that there was no long-term, consistent strategy guiding the import of technology. Introduction of technology is mainly through buying equipment or whole factories, which is often influenced by political environments. Due to the lack of a long-term, consistent policy, introduction of foreign technology did not make great contribution on a general scale to the development of China's economy as it should and it resulted in a situation of two extremes in industrial production: a very few factories equipped with sophisticated technologies and a large number of factories with old and backward equipment.

3.2 Failure to Realize the Software Components of Technology

When talking about technology, most people think of factories, machines, products or infrastructures (roads, communication systems, water distribution systems etc.) – the so called "hardwares". However, technology goes much beyond the hardware, and also comprises what can be called, by an analogy borrowed from computer engineering, the "software". This includes such non-material things as know-how, experience, education, organizational forms and so on.

The modern highly industrialized societies owe their development not merely to the invention and widespread application of new types of machinery from the steam engine of the first industrial revolution to the fifth generation computer of today, but also to major innovations and gradual improvements in organizational forms, institutional structures, management skills and educational levels.

The importance of these non-material innovations and improvements should never be underrated. Nicolas Jequier indicated that:

One of the major innovations in organizational forms in the first half of last century was legal invention, that of the limited company. This new form of association allowed potential entrepreneurs to escape from the stifling restrictions of the professional guilds inherited from the Middle Ages. It also consecrated the dismantling of the King's monopoly on industrial and commercial entrepreneurship, which had been institutionalized over the centuries by the system of royal charters. In fact, had it not been for the invention of the limited company, which released the entrepreneurial drive of a society in transition, the industrial revolution may well have aborted despite all the inventiveness in hardware.[16]

Another example is venture capital investment, which is primarily regarded as the early stage financing of relatively small, rapidly growing companies. This financial support sometimes is of key importance in deciding the life of an innovation in technology development. In an OTA report about technology, innovation and regional economic development, three key characteristics of venture capital investment were listed:

- It involves some potential equity participation for the venture capitalist, either through direct purchase of stock or through warrants, options, or convertible securities.
- It is a long-term investment discipline that often requires a period of 5 to 10 years for investment to provide a significant return.
- Venture capitalists are active, ongoing participants whose experience and specialized skills add value to their investment in a developing business; they are not passive investors offering only capital.[17]

In a sense, therefore, the venture capital industry institutionalizes many functions and benefits of the technological infrastructure of an entrepreneurial network.

When talking about know-how and experience, most of them must have been documented in some form. This can be a note for internal use, a patent application, or a publication. It is interesting to notice that an investigation of 210 cases of innovations in the energy and allied engineering areas in three countries (UK, FRG, and Sweden) showed that in 40 percent of all cases the new information was published (i.e. available publicly to anyone). While so large a percent of technologies is accessible to anyone in a country like the U.S., free of charge in specialized journals, they are sold as part of a package by commercial consultants to LDCs because this information is not available in those countries. There people tend to think hardware like factories, machines, products is more valuable than technical journals. Denis Goulet give us a very good example. He says:

A few years ago I met a Moroccan cement manufacturer. This man had worked for the government, but later left government service to create a company of his own in partnership with others. They built the first modern portland cement factory in Morocco.

He lamented that about two-thirds of start-up costs went to consultant firms for site studies, engineering studies and technology studies. In his conversation with me he declared had he known then what he knew ten years later just after having attended a certain number of U.N. conferences on technology and pre-investment studies – that a lot of non-proprietary technology was available in journals or in data banks of subsidized organizations like UNIDO – he could have reduced his preconstruction costs by about two-thirds.[18]

Today the range of new hardware which is available to LDCs as a result of the industrial research undertaken in the DCs is so wide and increasing so rapidly that it could, in theory if not in practice, meet a large part of their immediate needs. What is really lacking is the software we mentioned above and this is perhaps the area where the appropriate technology has the most to contribute. Hardware and the technical ability to produce it in an imitative way can generally be transferred from one country or culture to another. Organizational forms and social values are, by contrast, much more culture-specific and hence generally more difficult to transpose from one society to another. This has been clearly revealed in China's contemporary economic reform. How to introduce advanced technologies from Western countries without being influenced by their values has become the most perplexing problem yet to be solved by Chinese leaders.

3.3 Imperfect Market Conditions

It had long been believed that there are considerable advantages to being a "late-comer" in terms of industrial and technical development. The late-comers may make use of the vast accumulation of scientific and technological knowledge for which the DCs had to devote considerable resources.

But more and more LDCs have begun to complain that many transactions in technology take place under very imperfect market conditions in which the bargaining advantage is very much on the side of the enterprise selling the technology and hence the transfer process is costly to the buyer and involves considerable payment in foreign exchange.

Dr. P. Olpadwala of Cornell University did a thorough analysis on this aspect in his "Appropriate Technology For Forest Industries":

The market of technology, is now commonly accepted as being imperfect, especially in the international context. Large businesses with considerable market power tend to use the highly-developed and sophisticated legal system of international patents and trademarks to systematically exploit their technological resources. More often than not, this entails the withholding of technical knowledge as much as its diffusion. Further, when transfers (sales) do take place, these are accompanied by a multitude of problems and drawbacks from the LDCs buyers' perspective.[19]

The first and most obvious disadvantage, he maintained, pertains to cost. The highly imperfect nature of the market enables sellers to charge substantial premiums or rent for the technology. The most direct way of doing this is by overcharging, but some indirect methods are also used. Among them the most prominent one is what has been called the practice of bundling. Under this device, the essential requirements of the buyer are only met if the latter also agrees to purchase from the seller additional items which had not been requested. Sellers are able therefore to translate their market advantages into a bundle of things to be sold, including but also much exceeding the actual requirement of the buyer. This "bundle" could contain other technologies and/or products that the buyer does not need or ask for, and it could also force an administrative or financial arrangement on the buyer that is not in the latter's interest. The most common examples of the latter is foreign direct investment.

A second disadvantage is the restrictions imposed on the use of the technology for buyers of the technology. For example, buyers of international technology are often prohibited from exporting products made with these technologies, or at least restrained from directly challenging the seller's own markets in third countries. Restrictions may also be placed on buyers not to adjust, amend, or in some cases, even further develop the technology without the specific approval of the sellers. In all such instances the buyer is penalized, often heavily, and prospects for the local development of a more appropriate technology are also diminished.

Finally, imperfect markets for technology force LDCs to accept other deleterious and inappropriate features. For example, there is a need in LDCs to avoid redundancy in technology imports. While there are obviously many technologies worth importing initially or once, even at some cost, the best of these tend to lose their potential for genuine contribution to development if they are indiscriminately over-imported. Yet a seller's market can and does force weak buyers to repeat or renew technology transfer agreements beyond their true need, and in addition forces multiple concurrent imports of the same technology. It also compels buyers to accept an inflow of technologies for commodities which might otherwise by most reasonable standards be counted as unnecessary or non-essential for LDCs. All these add to the aggregate level of technological inappropriateness in LDCs.

4. HOW TO CHANGE THE SITUATION – THE EXAMPLE OF CHINA'S ENERGY INDUSTRY

Most efforts since the 1940s to reduce the gap between the world's rich and poor was called the Second Wave Strategy by Alvin Toffler in his popular book *The Third Wave*. He used Iran as a dramatic case in point. After describing the Iranian Shah's effort to make Iran the most advanced industrial country in the Middle East, Toffler said:

Nurtured by the West, attempting to apply the second wave strategy, the millionaires, generals, and hired technocrats who ran Teheran government conceived of development as a basically economic process. Religion, culture, family life, sexual roles – all these would take care of themselves if only the dollar signs were got right. Cultural authenticity meant little because, steeped in industry-reality, they saw the world as increasingly standardized rather than moving toward diversity. Resistance to Western ideas was simply dismissed as 'backward' by a cabinet 90 percent of whose members had been educated at Harvard, Berkeley, or European universities.[20]

Seeing the collapse of the Shah's regime in Teheran and the failure of the effort, Toffler asked "Is classical industrialization the only path to progress? And does it make any sense to imitate the industrial model at a time when industrial civilization itself is caught in its terminal agonies?" These questions are never easy to answer although many efforts have been made. In the following part let us take China's energy industry as an example to try to find some ways to solve the problems that LDCs face in transferring technologies from DCs.

4.1. General Situation of China's Energy Industry

4.1.1. Petroleum

Most of China's oil exploration has taken place in the northeast corridor, and that is where the giant fields and 75 percent of the reserves are located. Proved and probable reserves in the northeast are estimated at 10 to 15 billion barrels. Cumulative production has been 10 billion barrels, and additional discoveries and advanced technology may add an equivalent amount of oil resources.

Crude oil production rose rapidly in the 1970s to over 2 billion barrels per day in 1979 (the United States produces 8.6 million). By 1980, however, production flattened out due to the maturing of the biggest fields and a rapid decline in one unusual formation. Output in most fields has now stabilized or is increasing due to water injection to maintain pressure.

This situation has developed largely because of a lack of exploration. The fields in the northeast appeared to be so large that little attention was paid to the rate of depletion and the need to develop new fields. Even now that the situation has been recognized, China is spending only about $2 to $3 billion on oil exploration and development, about what a moderate size U.S. company would invest for a tiny fraction of China's production.

This has led to an extensive reevaluation of petroleum policies in China. The Ministry of Petroleum Industry is allowing foreign oil companies to participate in offshore exploration. More recently, as the focus of exploration has shifted to the northwest, which has a huge potential but harsh conditions, China has sought help from foreign companies in seismic surveys and exploratory drilling.

4.1.2. Natural gas

China's known natural gas reserves are only 4.6 trillion cubic feet, a much lower energy resource than crude oil reserves. However, gas has been a lower priority fuel because it is difficult to transport without an expensive pipeline system, and there is little export market. However, recent offshore drilling in the South China Sea has resulted in a commercially exploitable finding in the range of 3 to 7 trillion cubic feet. Basins in western China also show promise of very significant gas resources. So far natural gas production is limited to one basin in Sichuan and as a byproduct at the major oil fields.

4.1.3. Coal

China's coal deposits are gigantic, over 750 billion tons, and may be double that. Actual recoverable reserves are much lower, but China is in the same class with the U.S. or the U.S.S.R. At the present rate of exploitation, this coal would last hundreds of years. Most of it is reasonably good quality bituminous grade. Pockets of coal occur practically everywhere in the country, but the major seams are in the central and north central regions, far from the major industrial regions near the coast.

Currently, coal represents 74 percent of China's energy production. While the share is dropping, actual production has risen fairly steadily at an annual rate of 7 to 8 percent. The stated target for 2000 is 1.2 billion tons, about double the rate in 1980. Almost half of this goal would be met by small, local mines, but annual additions to large mine capacity will have to be 10 to 12 million tons per year.

Very little coal (about 10 percent of output) is cleaned before shipment. In many mines, non-combustible matter significantly increases shipping costs and causes problems in boilers when the coal is burned. Transportation is a major bottleneck. Coal already represents about 40 percent of all rail shipments.

4.1.4 Electricity
China's hydropower resources are huge, potentially as much as 380,000 megawatts (MW). Only 22,000 MW have been exploited. About 60 percent of the potential is in the Southwest. The electric power industry has been growing rapidly. Present total capacity is 81,000 MW, of which 68 percent is from steam plants and the rest is hydropower. There are six major regional grids and many small local grids. Twenty-two long-distance, high-voltage transmission lines have been built. There are 18 large hydropower stations and another 11 under construction for completion by 1990. The largest potential project, the "Three Gorges" on the Yangtze River, could produce 12,000 MW, but it is still in the planning phase and recently has evoked much controversy. This project, estimated at $9 to $12 billion would rank among the world's largest construction projects.

4.1.5. Nuclear Power
Nuclear power is relatively new in China. The only nuclear power plant under construction is the 300 MW Qingshan plant, 126 kilometers from Shanghai. Another larger one is to be built at Daya Bay in Guangdong province near Hong Kong. Sites have been chosen for two follow-on projects in Jiangsu and Liaoning provinces. Proposals for the former are being considered.

4.1.6. Energy Use

China's use of energy is quite inefficient. Artificially low energy prices and a shortage of capital have resulted in a vast amount of equipment and processes that were not designed to minimize energy use. It is now clear that demand for energy services will increase rapidly as the economy and standard of living rise, but that producing great amounts of additional energy will be very expensive, polluting, and in some cases, impossible. Therefore, to meet economic goals, increasing the efficiency of energy use will be necessary.

In the industrial sector which uses 72 percent of China's total primary energy, only about 40 percent of the energy is converted to useful service. The commercial/residential sector uses only 14 percent of China's total commercial energy. Coal is the major fuel for cooking and heating. In rural areas, noncommercial fuels (wood, crop byproducts, biogas) are very important. About one-third of the peasants have no electricity. It is a national goal to electrify all rural villages by 2000.

4.2. Suggestions on Technology in China's Energy Industry

4.2.1. A long-term and consistent strategy in developing energy industry must be made and be used to guide the practice of technology introduction.

Based on the situation of China's energy industry discussed above, we see that in almost every sector of the energy industry there is a need for new and advanced technologies, and some technology transfer has been going on. For instance, the offshore oil exploration projects have stimulated petroleum technology transfer to China, including training, joint technical services, and joint management. Many of the smaller U.S. oil field service companies are now participating. Another example is the nuclear power plant to be built in Daya Bay in Guangdong Province. All important components are going to be imported. The nuclear island (reactor, primary pumps, and steam generators) is expected to come from France, and the generating components from Great Britain. All these transferred technologies seem badly needed and appropriate from each sector's point of view. But if we look at the position of the whole energy industry, things are not so simple.

One problem we should be aware of is that of finance. Although compared to other LDCs China is in a relatively favorable position in terms of its foreign exchange holding, its energy needs are so great that it is still difficult to find the necessary resources. Foreign exchange reserves could be dissipated quickly with major purchases (e.g. nuclear power plants), and the uncertainties of the export potential of the energy industry for the remainder of the century in the face of rising domestic demand induces caution in the use of foreign exchange. Due to this financial constraint China cannot invest in all sectors of the energy industry to transfer advanced technologies on a large scale. Some appropriate choices must be made. Only based on a long term energy policy can the right choices be made.

Also the change in the structure of the energy industry will affect structures in other related industries. For example, in the 1950s the emphasis of China's energy industry was on coal; in the 1960s and until the mid-1970s it shifted to petroleum. Because of this change many new industrial boilers were designed to use oil as fuel, and what's more, many old industrial boilers originally based on using coal as fuel were changed to use oil. All this led to the drastic increase in the consumption of oil. In the late 1970s oil production proved to be far behind the need. To ease the situation, lots of boilers had to be changed back to use coal and some chemical plants based on oil as a raw material had to be closed temporarily. The estimated loss due to the switch of energy strategy was as high as several billion Yuan.[21] So a long-term and consistent strategy in energy development is needed not only for the development of the energy industry but also for the development of other industries.

4.2.2. Technology transfer should be combined with indigenous research and development

Although technology transfer from DCs can help LDCs to leapfrog the big difference in technology between DCs and LDCs, there are several reasons why this transfer should be combined with indigenous R&D.

First, even to operate imported technology efficiently requires considerable technological adaptation. The technology has to be adapted to local conditions – to differences in the quality/availability of various factors, in government relations, market conditions and so on. The more sophisticated the technology, the greater the need for technological resources, for effective adaptation and use. Japanese development in the twentieth century has been characterized by heavy use of Western technology. But Japan has also expended considerable resources on adapting this technology. One-third of its R&D has been devoted to adapting foreign technology, and the average expenditure devoted to adapting a unit of technology has been greater than the average expenditure on creating a unit of local technology.[22]

Second, so long as LDCs lack the capacity to create their own technological innovation, they remain dependent on the advanced countries for technology. This in itself is obviously an aspect of dependence. China's dependence on the Soviet Union for technology in the 1950s and the withdrawal of the Russians in the early 1960s is an example of the kind of result this dependence could lead to. The dependence of technology also puts LDCs in a weak bargaining position for the acquisition of technology. The market for technology is an area where bargaining necessarily plays an important role and there is considerable scope for bargaining. For the buyer the bargain is a difficult one because he does not know precisely what he is getting – if he did there would be no need to buy it. But the capacity to bargain generally depends on how much the buyer knows about the technology, about alternative sources of it, and how easily he could reproduce the technology himself. All these factors depend on the technological accumulation in the country concerned – its technological capacity. Thus, the greater the dependence, the worse a bargain likely to be struck and the higher the cost to the purchaser.

In contrast to many LDCs, China has an established energy R&D network. Several universities are especially devoted to training scientists and engineers for the energy industry. Thus, in the energy area, almost all sectors have research, design, and educational institutions which have more than 25 years of experience. Many of these had experience with technology transfers from the Soviet Union in the 1950s, and experience with technological self-reliance since 1960. This is a significant resource that China should use in indigenous R&D to develop its energy industry and to assimilate foreign technology.

4.2.3. Introduction of hardware should be combined with introduction of software

As we have mentioned earlier, failure to realize the importance of software components of technology is one of the reasons for the inappropriateness of technology transferred by LDCs from DCs. This software includes such non-material things as experience, know-how, management skills, education and organizational forms. In the area of energy, an outstanding example has been the Tennessee Valley Authority, one of the largest energy resource management organizations in the United States.

The most remarkable thing about TVA is not the power it produces, nor its accomplishment in flood control and navigation, nor the great change in the economic and social well-being of the people living in the river basin – but all these together. The way these were achieved and the thinking behind these accomplishments were well expressed almost thirty years before the creation of TVA by Gifford Pinchot:

Suddenly, the idea flashed through my head that there was a unity in this complication – that the relation of one resource to another was not the end of the story. Here were no longer a lot of different, independent, and often antagonistic thinking. In place of them, here was one single question with many parts. Seen in this new light, all these separate questions fitted into and made up the one great central problem of the use of the earth for the good of man.[23]

The approach, by which TVA's achievement was made, is called today "integrated regional resources management" and the experiences of TVA, now more than 50 years in existence, offer an illustration of how integration can be achieved and how it can benefit a region, the resources, and the people.

When the first managers began assembling in the Tennessee Valley, they began to grasp the complexity of the problems. At first glance it seemed the only resource the valley region offered was people – relatively untrained people who were moving out of the area to get jobs in the crowded cities to the north. Once they started more comprehensive analyses of available resources of the region and their capability to provide benefits needed to enhance the Valley's future, they saw the need for a comprehensive program to upgrade the current farming practices in the agricultural resources. If that were accomplished, the contribution of the agribusiness sector to the overall economy could be greatly enhanced. Likewise, the forest could be revitalized with plantations of fast-growing pine which, in the short term, would help slow erosion and also provide a potentially valuable raw material for wood industry expansion over the longer term. With dams to control floods, the river flow would be more constant and river transport could be expanded. Also, if dams were needed for flood control and navigation, they could also produce power, which could be used to make the area more attractive for industry.

As various experts joined together to study the problems and opportunities, they increasingly found that one expert's "problem" offered another expert an opportunity. Planning together, agriculturalists, engineers, and social scientists were able to map a strategy that called for taking people off the farms and giving them enough training to assist in the construction process. They would develop a variety of skills; and when the work was completed, these people would constitute a trained work force that would become a source of labor for industrial customers attracted to the Valley by the new power system.

Individually they probably would not have envisioned or mapped such a complex strategy. Together they found unexpected solutions and opportunities. Although the terms "system approach" or "integrated regional resource management" were not in common use at that time, their practice is probably one of the most effective demonstrations of regional development known in the world.

If we take a look at China's underdeveloped areas and its vast untapped resources, especially hydropower resources in those areas, we will understand how valuable TVA's experience is for the development of those areas. They do need sophisticated technology, but what they need more is to learn the way in which TVA tackled all its difficulties and the experience TVA has gained in more than 50 years. We are glad to see that in recent years cooperation between TVA and China has begun in the integrated resource development of the Hongshui River area in Guangxi Zhuang Autonomous Region in South China.

Certainly one can not underestimate the difficulties people are confronted with in effecting technology transfer, especially the software aspect. But witnessing the talents human beings have displayed in creating technological wonders we have every reason to believe that people will be able to manifest their talents too in using these wonders for the well-being of mankind.

REFERENCES

Bernal, J.D. *The Social Function of Science*, Routledge & Kegan Paul Ltd, London, 1939.
Bhallasali, B.N. *Transfer of Technology Among the Developing Countries*, Asian Production Organization, Tokyo, 1972.
Dorf, Richard C. and Hunter, Yvonne L., (ed.), *Appropriate Visions–Technology, the Environment and the Individual*, Boyd & Fraser Publird S. *Appropriate Technologies for Developing Countries*, National Academy of Science, Washington D.C., 1977.
Ho, Samuel P.S. and Huenemann, Ralph W. *China's Open Door Policy–The Quest for*

Foreign Technology and Capital, University of British Columbia press, Vancouver, Canada, 1984.

Jequier, Nicolas. *Appropriate Technology–Problems and Promiscs*, OECD, Paris, 1976.

Reynolds, Lloyd G. *Economic Growth in the Third World: An Introduction*, New Haven, Yale University Press, 1985.

Schumacher, E.F. *Small is Beautiful*, Harper & Row, New York, 1973.

Uyehara, Cecil H. *Technological Exchange: The U.S.–Japanese Experience*, University Press of America, Inc. Washington D.C., 1982.

Technology, Innovation and Regional Economic Development, Washington D.C., U.S. Congress, Office of Technology Assessment, OTA–STI–238, July, 1984.

NOTES

[1] J.D. Bernal, *The Social Function of Science*, MIT Press, 1967. p. xiv.

[2] J. Schmookler, "The Changing Efficiency of the American Economy 1869–1938", *Review of Economic and Statistics*, vol. 34, August 1952. pp. 214–231.

[3] M. Abramovitz, "Resources and Output Trends in the United States since 1870" *American Economic Association Papers and Proceedings*, vol. 46, May 1956, pp. 5–23. Kuznets, *National Product Since 1869*, New York, National Bureau of Economic Research, 1946.

[4] See W.P. Hogan, "Technical Progress and Production Function", *Review of Economics and Statistics*, vol. 40, Nov. 1958, pp. 407–411.

[5] See J. Annerstadt, *Technological Dependence: A permanent phenomenon of world inequality*, mimeo, Institute of Social and Economic Planning, Roskilde University Center, 1978.

[6] See UNCTAD, *The Role of the Patent System in the Transfer of Technology to Developing Countries*, TD/B/AC.11/19, Rev. 1, 1975.

[7] See Nicholas Jequier, ed., *Appropriate Technology: Problem and Promises*, Chapter I, "The Origin and Meaning of Appropriate Technology".

[8] Nicolas Jequier, p. 21.

[9] See Sim Van Der Ryn, "What is Appropriate Technology?", *Appropriate Visions*, Edited by Richard C. Dorf & Yvonne L. Hunter, Boyd & Praser Publishing Company, San Francisco.

[10] See Richard S. Eckaus, "Alternative Criteria of Appropriateness of Technological Decisions", *Appropriate Technologies for Developing Countries*, National Academy of Sciences, Washington D.C., 1977.

[11] See Dr. James C. Abegglen "U.S.–Japan Technological Exchange in Retrospect, 1946–1981", *Technological Exchange: The U.S.–Japan Experience*, Proceedings of a Symposium held on Oct. 21, 1981, edited by Cecil H. Uyehara. Also see Samuel P.S. Ho and Ralph W. Huenemann, "Chapter 1. China's Development Strategy and the Open Door", *China's Open Door Policy – The Quest for Foreign Technology and Capital*, University of British Columbia Press, Vancouver, 1984.

[12] Samuel P.S. Ho and Ralph W. Huenemann, p. 14.

[13] Chen Huiqin, "The Orientation of Technology Transfer Must Be Changed", *Jinggi Guanli [Economic Management]* April 1981, p. 22.

[14] *Ibid.*, pp. 22–23.

[15] See Cao Jiarui, "The Contemporary Situation and Problems in China's Technology Introduction", *Outlook Weekly*, No. 18, May 15, 1986, p. 14.

[16] Nicolas Jequier, pp. 21–22.

[17] See *Technology, Innovation and Regional Economic Development*, OTA Report July 1984, Washington, D.C., p. 41.

[18] Denis Goulet, "Value Orientation in Technology Policy", *Bulletin of Science, Technology and Society*, vol. 3, p. 306.

[19] P. Olpadwala, "Appropriate Technology for Forest Industries", Paper prepared for the Forest Department, Food and Agriculture Organization of the United Nations, April 1985, p. 24.

[20] Alvin Toffler, *The Third Wave*, William Morrow and Company, New York, 1980, p. 347.

[21] See Liu Guoguang, "Chapter 6: The Energy Problem in Chinese Economic Development" *Research on Strategic Problems of China's Economic Development*, Shanghai People's Publishing House, Shanghai, 1984.

[22] T. Ozawa, *Imitation, Innovation and Trade: A Study of Foreign Licensing Operations in Japan*, Ph.D. dissertation, Columbia University, 1966.

[23] Gifford Pinchot, *Breaking New Ground*, Harcourt, Brace and Company, New York, 1947.

S. MUTHUCHIDAMBARAM

DIFFUSION OF TECHNOLOGY VIS-A-VIS TRANSFORMATION:
INCREASING CONTRADICTIONS BETWEEN TECHNOCRATIC MARKET VALUES AND SOCIAL DEMOCRATIC VALUES

INTRODUCTION

The world at present is undergoing various crises and among them a set of twin-crises stands out: a crisis of technology and a crisis of democracy. These two crises are linked and reinforce each other at a global level. The value premises behind these concomitant variables are not functionally and normatively synchronized. The purpose of this paper is to identify the contradictions between the technocratic market values (productivist logic) and the social democratic values (logic of needs) within the advanced countries (core-countries) and to project and consider these contradictions with reference to transfer of technology to the Third World Countries (TWCs/peripheral countries) through a paradigm.

PARADIGM

The paradigm given below is just a pattern or framework to give organization and direction to the subject matter under consideration. It is descriptive and no ontologic-predictive element is intended in it. Since we deal with "contradictions" between two sets of values it contains an element of dialectical paradigm.[1] [p. 228]

TERMINOLOGY AND CONCEPTS

"Technology" and *"technique"* are used interchangeably. They mean the totality of methods rationally arrived at and having efficiency as an

227

Edmund F. Byrne and Joseph C. Pitt (eds.), Technological Transformation:
Contextual and Conceptual Implications, 227–248.
© 1989 Kluwer Academic Publishers.

Characteristics of two systems of social production and technique[2]

	Productivist logic	Logic of needs
Objectives	Enterprises seek to maximize their profits, the structure of consumption is a dependent variable. The state seeks to maximize in the long term the volume of consumption; the structure of consumption is largely dependent on the demonstration effect.	Seeking to satisfy the concrete needs of the population, particularly the poorest classes; seeking appropriate solutions on the basis of social missions to be performed.
Role of planning	Correction of the most glaring dysfunctions; creation of the socioeconomic environment favourable to the action of the enterprise. Organization of productive activities in the light of productivist logic; planning is subordinated to the economy.	Identification of societal objectives and organization of social activities with a view to their implementation; the operating of the economy is subject to planning; the place ascribed to formal optimization procedures is reduced in favour of largely participatory processes.
Time horizon	Only the short term and the medium term are taken into account.	Deciding between the short term and the long term is the planner's central concern.
Criterion of rationality in terms of:		
Resources	From the angle of availability and price, wastefulness by definition does not exist, and the husbanding of resources is proposed.	Conservation and development of resources from a sense of solidarity with future generations, husbanding of resources, and struggle against wastefulness.
Space	From the angle of availability and price, space is regarded as an ordinary resource.	Guarding of options for the future and harmonizing of the many purposes space may have.
State of the environment	Correction of negative external effects is harmful to production.	Prevention of harmful phenomena, and respect for long-term ecological equilibria.

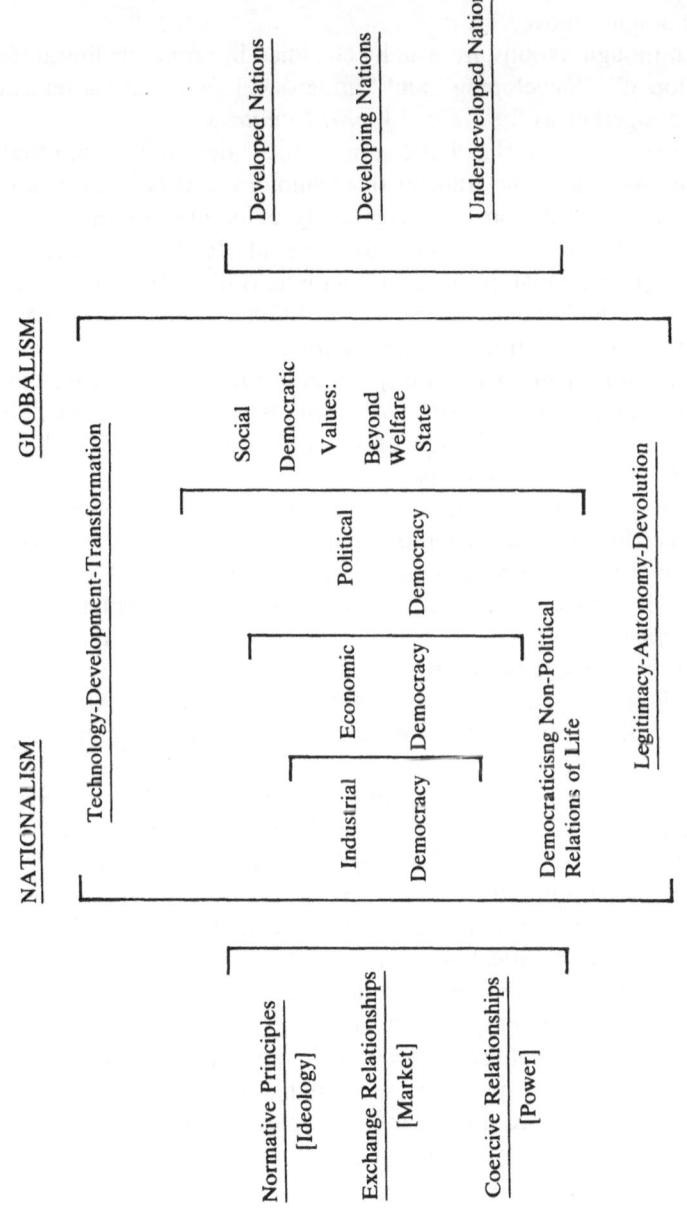

absolute goal for a given stage of development in every field of human activity. This dominant model is identified under "productivist logic" in the paradigm, above.

Even though nations are usually classified in economic literature into "developed", "developing" and "underdeveloped", the latter two are lumped together as the *"Third World Countries"*.

"Nation-state" is the basic unit which determines international relations as well as the transfer of technology between and among the nation-states. A nation is a large body of people possessing its own territory ruled under a unified government. It also refers to *"peculiarities"*, to national pride and "nationalism". The latter means a devotion to the interests of one's own nation as well as the advocacy of or movement for national advancement or independence.

On the other hand, when we speak of technological transfer/diffusion we are speaking in terms of *"globalism"* which is an idea or policy of placing the interests of the entire world (global village) above the interests of particular nations.

This is where the question of "nationalism" and "globalism" come into focus in terms of technological transfer and the inherent conflicts between two sets of values, namely, technology-growth-development vis-à-vis legitimacy, autonomy, devolution and transformation.

Transformation is not a simple transfer, in the sense that we are to remove from one situation to another. On the other hand, it means to *change in form and structure*, meaning to change in *condition, nature*, or *character*. In that sense, transformation is a synonym for metamorphosis or transfiguration.

Transformation is a function of synchronizing certain organizational principles within and between nations.[3] These principles are the following: *normative structures and principles* which contain the basic frame of references and values or what is known as *legitimation system*; *exchange relationships* determine the economic system; and the *coercive relationships* denote the political system. These three organizational principles shape social production and technique, exchange, political choice, bureaucracy and bargaining system within and between nations.[4] In terms of transfer of technology and economic development, the controversy revolves around the questions of exchange principles (market) dominating the other two principles of organization namely, normative principles and political system.

Is the "invisible hand" a necessary and sufficient condition to achieve justice and equality in the social production and distribution? If it is not, is there a need for democratising non-political relations of life?

In the latter context, *industrial democracy* denotes a social and economic system in which labour is at least partially substituted for management by institutionalizing the rights of workers and their representatives to have an influence on and humanize the decision-making process in the firm in which they employed. This requires a change not only in the productive forces (resources and technology) but also in the productive relations (human relations), transcending the market oriented relationships.[5]

This transformation may not take place unless the principles of democratization are applied at the macro-economic level where there is high concentration of power, control, and authority. In this sense, *economic democracy* denotes a system which would introduce democratic principles into decision-making in the economic sphere and in the distribution of power and control in the political economy as a whole. It is inappropriate that far-reaching economic and social-political decisions should be left to a few individuals, who in the main represent the private interests of a minority.[6]

In the following pages the above paradigm and concepts are applied to substantiate the theme of this paper.

TECHNOLOGICAL DIFFUSION AMONG THE ADVANCED COUNTRIES

The underlying philosophy behind technological diffusion among the advanced countries (ACs) is the Schumpeterian "process of Creative Destruction", where we "all" advance at the expense of the "few" and the destruction in the process of creation is not only necessary but inevitable. That, according to Schumpeter, is the *essential fact* about innovations under capitalism.[7]

Since the mid-1970s, technical-industrial modernization through new technology (microelectronics) among the ACs has been taking place in a *mindless and chaotic manner*. This process of modernization has no conception of the relationship between destruction and construction, modernization and liberation – a relationship that today can no longer be considered simply in "productivist" terms.[8] While the imperative of modernization has lost its perspective, it has also dramatically increased the relative significance of those people, institutions, values, and conditions of life that bear the cost of modernization.

Policy concerning technology, education, family life, economic growth neither checks the process of modernization nor guides it towards socially meaningful goals. Nowhere has democratization been accomplished to the point where the private character of decisions concerning the volume, kind, point in time, location of investment and transfer of technology has been effectively transformed in the manner of democratic control.[9] On the other hand the following factors have cumulatively strengthened the productivist-market values at' the expense of logic of needs and social democratic values:

– the kind and nature of technological change
– global economic crisis
– globalization of world economy by industries
– the rise of Transnational Corporations (TNCs), oligopolistic competition, global demand management and transfers
– decline of Keynesian Welfare State (KWS)
– revival of neo-laissez-faire ideology (NLFI)
– rise of monetarism and militarism and
– globalization of local conflicts by super powers

Since it will be beyond the scope and space assigned for this paper, these contextual factors are only briefly explored.

The new technology (NT), as distinct from the previous forms of technologies, is characterized by the following attributes of such a magnitude that they have brought about, in effect, a crisis of technology: enormous reduction in size and cost; increased speed and reliability; improved energy efficiency and portability and ten times faster diffusion rate than that of any previous technology.[10] "The difference in degree of speed of microelectronic product and process innovations, in the speed of their diffusion, and in the breadth of their applications compared with earlier major technological developments may amount to a *difference in kinds*".[11]

While the period of automation since World War II coincided with a period of prosperity, the NT diffusion has been taking place in an environment of global economic crisis, and globalization of world economy by industries from the ACs through the rise and spread of TNCs, oligopolistic competition, global demand management and transfers.

Production of machines themselves was relatively labour intensive during the earlier period; and impact of automation was primarily on the blue collar labour market. The economic prosperity combined with the active role played by KWS had contributed to the extensive growth of labour intensive service sector which in turn more than compensated for the decline in blue collar jobs and created an environment of near full employment, economic security and growth.

On the other hand, NT is contemporaneous with the decline of KWS, and the revival of NLFI which in turn has resuscitated the Smithian market, monetarism, and Social Darwinism, within and between nations. The impact of NT, under these circumstances, is far reaching. Unlike automation, the NT has penetrated into all sectors of labour market, industry, and institutions. The service sector, which served as a buffer in neutralizing the impact of structural change in the previous period, is itself the main target of the NT. While Taylorism and Fordism – standardization of people to meet the imperatives of the technique to achieve efficiency – was previously applied primarily in the manufacturing sector, with the advent of NT, they have become the universal principle applicable to all sectors of the political economy.

Reaganomics in the U.S. and Thatcherism in England are the epitome of NLFI which in fact has penetrated even into countries with social democratic leanings (e.g. West Germany and France). This ideology contains contradictory public policy components. It actively promotes technological change with productivist and Schumpeterian logic. On the other hand it acts as a demolition-derby through other policies, namely: denationalization, deregulation, privatization, contracting-out public service, either non-enforcement, or dilution or repeal of occupational health and safety protection, deregulation of plant closures, shifting the balance of power in favour of employers and against unions, promotion of deunionization and encouragement of a union-free environment, and weakening of the social security system.[12]

These structural changes and monetarism of NLFI are taking place along with militarism and all of them are based on the "trickle-down theory" which assumes that in the long run the benefits from these three components spill over to the whole society.

In 1980 the world military expenditure was $450 billion and by 1982, it was $650 billion.[13] In the recent one-trillion dollars budget of the U.S., one-third ($312 billion) is allocated for defense purposes, an increase of three percent.[14] The current military expenditures around the world are estimated to be $20 billion every ten days.[15]

In dealing with North-South co-operation for world development, the Brandt Commission has identified the contradictions in monetarism-militarism – trickle-down-theory in the North in these words:[16]

. . . One of the tragic developments of the last three years has been the rise in arms spending. Some think that this will help the world out of recession. In fact military expenditure is very much more a part of the world's economic problem than its solution. At any given level of public expenditure, the higher the proportion of spending devoted to weapons procurement, the small the amount of employment created. Military expenditure may also be more inflationary than other public spending. *The alleged benefit of technological spin-off is also fallacious; technological advance can be promoted directly with far greater economy.*

There is no lack of political support for other forms of public expenditure: on the contrary, the public of the industrial North frequently complains of the cuts in social and infrastructure investment. It cannot be argued that these economies require high levels of military expenditure for any economic purpose. *And high military expenditure is an obstacle to the pursuit of all those objectives which affect their national security and to which military measures are not relevant: those objectives include the development of the Third World.* [Emphasis Added].

TECHNOLOGY TRANSFER (TT) AND THE THIRD WORLD COUNTRIES

Ideal Social Praxis for TT

Science and technology are both part of the problems and part of the solutions in the sense that "technology is simultaneously the bearer and the destroyer of precious human values, and that although it brings new freedom from old constraints imposed by nature, tradition, or ancient social patterns, technology also introduces new determinism into the life of its adepts".[17] This unity of opposites has been critically emphasized by the World Council of Churches by focusing on the power of science and technology to create and power to destroy, and the power to promise and the power to threaten. The Council summed up its views in these words:[18]

. . . Science through its contribution to understanding liberates people from many forms of ignorance and superstition. Technology liberates them from many physical constraints and insecurities . . . Yet in the fact of all these promises, science and technology appear to many people as threats . . . It is not only destructive human purposes that turn science and technology from promise to threat. Even well-intended uses of technology have unintended consequences that perplex or frustrate the people who initiated them.

The praxis suggested by the Council to avoid this predicament is *justice-participation sustainability.*

A similar praxis has been arrived at from the secular side in the Schevenigan Report, "Towards a New International Development Strategy", by singling out *participation and decentralization, demilitarization, respect for cultural alternatives,* and *sustainability* as principles of a social praxis oriented toward a *non-exploitative development path* in TT.[19]

THREE APPROACHES TO TECHNOLOGY TRANSFER (TT)

A Dependencia-Approach, a Global Problems Approach and a Social Control Approach are in vogue regarding TT[20] and each differs in various degrees from the social praxis identified above. These aproaches have different bearings on the following factors regarding TT: cost, control, suitability and local innovation.[21]

Dependencia-Approach is based on the market model of a liberal open-door policy toward TT to TWCs. Justification for this approach is that only the ACs have the necessary technology and R & D and that the TWCs do not have the necessary supply of indigenous scientific and technological capabilities and therefore what is needed is the importation of foreign technologies.

Given the fact that almost all the TWCs, until the recent past, were either former colonies or satellites of some of the ACs, this approach is perceived to be a form of neo-colonialism based on technological hegemony of the ACs. Also the TWCs are concerned about the transfer cost, loss of control over the nature and use of the imported technology in enterprises operating on their territories. Further, such a transfer may prevent the recipient country from determining the suitability of the technology through "unpacking" and "modification". All these factors in the dependencia-approach lead to inhibition of local innovation.

Global Problems Approach has been developed through North-South dialogue as exemplified in the Brandt Commission 1983, *Common Crisis North-South: Co-operation For World Recovery.*[22] It starts with the assertion that science and technology are both part of the problems and part of the solutions. Globe is treated as "spaceship earth". It emphasizes the dynamic imbalances that are developing between human activities and the socio-economic-political environment within which these activities are carried on. Since problems are global, the solutions must be sought on that basis.

The success of this approach depends upon the following: the effectiveness of North-South dialogue; shift from militarism to pacifism through the reduction of East-West conflict; and the effectiveness of the U.N. and its various agencies.

The gap between the normative values inherent in this approach and the concrete reality and realpolitik is not narrowing. The world still functions on the dictum that all nations are equal but some are more equal than others. The TWCs perceive themselves to be the "others".

Social Control Approach differs from the previous two in the identification of as well as solution to the problems of the TT to the TWCs. The following words of the 1975 Dag Hammarskjold Report exemplify this approach:[23] "Development of science and technology has become *primarily a political and social issue, not a technical one*. Producing technology, in the present international structure, means producing instruments of control and influencing over other individuals, firms and nations. *The capacity of technology to transform the nature, orientation and purpose of development is such that the question of who controls technology is central to who controls development . . .*"

Under the first approach any socio-economic issue is converted into a technical issue and hence the solution becomes a technocratic one; and further the dependence of many TWCs has been increased by technological solutions isolated from the economic, social and ecological context to which they apply. The Global Approach has taken "too big a bite", without the necessary infrastructure, "to swallow and digest" it. While the first is pure market model, the second is based on certain non-market normatives: the solutions offered by the latter focus on the common predicament of humanity rather than on the differences of access, by people or nations, to the goods and of exposure to the ills of the contemporary world.[24]

The Social Control Approach attempts to overcome these problems through the *social-democratic endogenization of science and technology rather than the merely national endogenization*. In the opinion of Volker Rittberger this approach is still in its embryonic form, neither approved by the leading ACs, nor practised by many TWCs:[25] "However, the merit of the social control approach is that, at the very least, it highlights the need for a political regime for science and technology, both at the national and at the international levels, that is capable of giving effective representation to diverse, even conflicting, interests concerning the collective uses of science and technology in developing as well as industrial developed countries".

Among the models described above, the Dependencia Approach remains the dominant one in TT to TWCs as reflected in the UN and its various agencies[26] and even more in the transnational corporations (TNCs) as carriers of TT.

TRANSNATIONAL CORPORATIONS AND TT

Transnational corporations (TNCs) are at present the most rapidly expanding institution which has introduced a revolutionary structural modification not only of the Western economic system but also of the economic system of the TWCs. It is revolutionary in the sense that in creating and integrating global economy, the TNCs are weakening the political position of the nation state and its ability to manage its own economy.[27]

The TWCs perceive the role of TNCs not as the transferrer of technology but as a transmission belt for neo-colonialism and development of underdevelopment through the integration of the global production, marketing and finance system dominated by the TNCs' headquarters in their home countries.[28] In this sense TNCs perpetuate the Dependencia model and prevent the Social Control Approach to TT for the following reasons:

TNCs act in the long run to maximize their interests, making the TWCs, as long as they are highly dependent actors in this system, very vulnerable. Integration of TWCs' economies into this system has a sharply distorting effect on those countries, socially and technologically:

Socially, through the now well-recognized phenomenon of the dual economy, in which 10–20 percent of the population participating in and served by the modern sector have the possibility of realizing some benefits from such integration, while the remaining 80–90 percent of the people stagnate in deprivation and misery.

Technologically, through reliance on external sources of technology, leading to failure to develop and use local problem-solving capacities and to trivialize traditional technologies, which are simply overwhelmed by "superior" foreign technologies.

The weakening political position of the TWC-nation state and its ability to manage its own economy (economic democracy in its elementary sense) in the light of TNCs role imply:[29]

– questioning the viability of regional economies or regional politico-economic communities;

- that balances of trade and payments calculated upon conventional aggregates have become meaningless (and even fictional) and beyond the control of governments;
- that new inflationary pressures and price movements are being generated by new corporate practices, especially long-term global self-financing, and are therefore unresponsive to accepted fiscal and monetary theory and policy;
- that the mobility and geocentric form of the multinational corporation naturally lends itself to utilization of tax havens and evasion of national fiscal and tax obligations, on which counteracting measures taken nationally can only have a limited effect at best.

In short, these developments undermine completely the conventional Smithian economic thinking embodied in Reaganomics and Thatcherism.

The other side of the coin is the question of industrial democracy as defined earlier. The TNCs increasingly *condition* the effectiveness and relationship between management (capital) and labour in collective bargaining (industrial democracy in its rudimentary form). "Capital's" advantages over "labour" are enhanced in the following ways:

- ability to locate investment in favourable circumstances relative to low wages, weak unions, favourable tax laws and hospitable pro-management legislation;
- transfer of production from one country to another to offset the economic effect of a strike;
- minimization of tax obligations and maximization of profit-positions where desired through adjustment of intercompany or transfer pricing;
- exploitation of advantages of tax havens – concentration of research and development in parental facilities to effect economy-of-scale advantages beyond the means of national companies;
- raising capital on world markets and freeing the company from subservience to national regulations on: tariffs, capital, cartels, taxes, publishing of financial statements and the like.

The social power as well as the blackmail capacity of TNCs (their ability to abstain from investing or "capital-strike") repels any chance for state intervention to enhance social control. The more advanced TWCs have become increasingly integrated technologically with the industrialized countries through TNCs. Whatever TT is introduced through the TNCs inevitably leads to "development against people and deindustrialization".[30]

The pattern of interlocking alliance of power, privilege, and technology between the technocratic elites in the ACs and those in the TWCs inevitably leads to development of underdevelopment and reinforces the dependencia approach.[31]

To the extent that governments consist of individuals who benefit from, the political economy in being, they may not wish or be able to challenge it. An alternative technology at a macro level involves an alternative political economy – a different distribution of the benefits of the economic system. Governments which have developed in one system may not be powerful enough to choose an alternative system . . .

The effective pursuit of an alternative appropriate technology would threaten interests in the advanced countries . . . who are currently benefiting from the use of advanced-country technology in developing countries. The continued use of advanced-country technology is at the heart of the continued dependence of the poor countries. It maintains the advanced countries' lead in technology and therefore permits, indeed necessitates, the continued sale of technology, goods and managerial services to poor countries, on terms favorable to the rich countries.

There is a *Code of Conduct* prescribed by the United Nations regarding transfer of technology to prevent the above situation[32] but the discrepancy between the actual conduct and the prescription is so wide that one may be tempted to say that the *Code* is a heroic attempt to convert a tiger into a vegetarian.

THE RISE OF "GROUP OF 77": A COUNTERVAILING FORCE

In the late 1970s, a "Group of 77" TWCs has come into existence within the framework of the United Nations.[33] In its purpose, the Group is a united front, countervailing force, and a bargaining mechanism for the developing countries in dealing with the ACs and TNCs. It is a kind of OECD for TWCs. It aims at getting away from Dependencia Approach with due appreciation for some of the normative elements in the Global Problems Approach:[34]

. . . The group system as a negotiating method has not been "invented" like penicillin or the aeroplane, but originated logically from a particular situation which has been created by the reappearance on the international scene of a great number of independent countries, formerly subject to colonial domination. Decolonization implied an essentially political struggle, with periods of violence and periods of negotiations. The "group" in the setting of the United Nations is the practical application of the only bargaining power of the economically weak, a political alliance, in direct continuation of the struggle for decolonization. It is undeniable that the group system has a great number of shortcomings as a negotiating tool. But in respect of the growing, though nevertheless somewhat faulty criticism towards it, one should ask a simple question: What is the alternative for the economically weak? The group system is the expression of that "labour movement style" solidarity, . . .

Superpowers want and expect the TWCs to follow in their footsteps in terms of basic ideologies and development technique. Neither the American model nor the Russian model is applicable to the TWCs. Therefore they opt for a Third or Fourth model or a hybrid model. These alternatives have already been experimented with by a few developed Scandinavian or Nordic countries in Europe with relative success.[35] The Group of 77 begins to assert the need for such a choice and experimentation. The offspring of this crossing is a breed of planning which is as different from the planning which has materialized in the Western countries as it is from Soviet planning.[36]

The ACs in general and the TNCs particularly consider the following as contentious issues regarding TT and economic development in the TWCs:

- the scope and intensity of state intervention in the process of promoting and applying science and technology to development.
- the methods of sharing in the world's scientific and technological potential; and
- the provision and allocation of financial resources for the development of science and technology in the developing countries in particular.

Both the ACs and the TNCs tend to insist on safeguarding the relative autonomy of the private sector both at home and abroad as a means of optimizing the growth and practical use of the available scientific-technological potential.[37] This insistence is perceived by the Group of 77 as "Global Integrationist" strategy towards TT.

In the 1980s, TWCs have opted for a "least worst" strategy of selective technological and economic delinking and disassociation with the ACs and TNCs to achieve the following objectives:[38]

- technological autonomy, not capacity, as the goal of developing countries
- selective technological disassociation as a strategy for achieving a greater autonomy;

– and technological autonomy (versus dependence) in achieving greater equity among and within nations.

Technological autonomy is not autarky but *greater selectivity* in and *closer control of externally acquired technology*. Selective disassociation, does not mean technological isolation forever but it does mean *carefully targeted technology acquisition on the initiative of developing countries*. This interim strategy does not prevent the TWCs subsequently relinking with the international system into a more equitable and autonomous relationship. The value of this strategy is summed up by Ward Morehouse in these words:[39]

There is nonetheless no clear-cut set of causal relationships that should lead us to conclude that greater technological autonomy will necessarily lead to greater social equity within countries, although few developing countries have yet been able to meet the minimum needs of most of their people on a self-sustaining basis without a substantial measure of technological autonomy. Indeed, technology is typically a dependent variable in the complex calculus of social change, and economic, political, and cultural factors matter much more. On the other hand, greater technological autonomy appears to be an indispensable condition in realizing greater equity among nations. Countries that have limited problem-solving capacities of their own and that are largely dependent on others for providing the knowledge and skills needed to solve their problems are always going to be vulnerable, in the long if not the short run.

THIRD WORLD MULTINATIONALS (TWMs) AND TT

In addition to the above strategy of selective delinking for achieving autonomy and social control over TT, the use of TWMs as a substitute for TNCs, has increased among the TWCs.[40]

The unilateral and supernational decision-making powers of the TNCs and potential for adverse industrial relations consequences is considerably greater in the TWCs with weak trade unions, a buyers' labour market, less well-entrenched industrial relations institutions and regulatory arrangements. With the increasing experience and understanding by the TWCs regarding this North-South TT problems, they have begun to explore the possibilities of using TWMs as a South-South co-operation mechanism.[41]

TWMs offer technology that contributes to host countries' *employment objectives*, *save on capital*, and *reduces bills for imports*. The most striking characteristic of technology transferred by TWMs is its *greater labour intensity* than TNCs from the ACs. TWMs typically transfer technology that is particularly suited to manufacture at small volumes[42] One of the important ways used by TWMs in developing countries to adapt to small size markets is to build *flexible plants*, setting them up so that they can produce a wide range of products or product models.[43] TWMs are more likely to *use local materials* than are TNCs from the advanced countries. While TNCs from the advanced countries are generally ethnocentric (home country-oriented) with some exceptions tending towards geocentrism (world-oriented), the TWMs are usually polycentric (host country-oriented).[44] Louis T. Wells, Jr. has made the following observation regarding TWMs.[45]

> Third World multinationals have a great deal of appeal, and not only on narrow economic grounds. They represent a significant and growing forum of south-south cooperation. They offer competition to the traditional multinational firms in certain industries, a rivalry that can lower the cost of technology. Moreover, they offer a way for a host country to diversify its sources of investment. By encouraging firms from other developing countries, a host country can loosen its dependency on a small number of industrialized countries.

In industrial relations matters, the TWMs are more flexible and accommodative to the local conditions and practices than the TNCs from the ACs. The unions in the host countries do not feel as much overwhelmed by the TWMs as they do in the case of the supernational giants from the ACs whose threat to withdraw the investment is the most potent weapon against not only the unions but also against the governments in the TWCs.

There is no guarantee that these strategies of selective delinking and substitution of TWMs for TNCs of ACs will be sustainable in the long run; but at least these are worthy attempts under the present circumstances of expanding economic and technological hegemony:[46]

> When the world had rebelled against the West and regained its freedom, when the soldiers and missionaries had departed, the West retained its power and continued its program of exploitation but by other means. In the process it has become ten times the hypocrite it was before. It controls the economies of the third world and keeps two-thirds of mankind hungry. Thanks to unjust agreements, to the laws of the international market, to a unilateral regulation of prices, and to the use of tariffs, it continues to steal the wealth of peoples who think of themselves as now free who are in fact financial and economic dependents.
> Multinational corporations enter like cancerous growths into the weak economies of these confused and traumatized nations. All the wealth still flows out to the West, even if now by different ways, and the natives gain nothing from it.

SHATTERED DREAMS OF JACQUES ELLUL

No one has written as much as Jacques Ellul did on technique and humanism based on Biblical faith and hope and with frequent pessimism and occasional optimism.[47] According to Ellul, technique is the root cause for economic poverty and political poverty; the latter refers to lack of participation in decision making regarding the nature and use of technique at micro and macro level within nations, particularly in the TWCs.

There was a time when Ellul thought some TWCs, such as Morocco, India, and China, with strongly rooted popular culture might be able to maintain their singularity while absorbing a limited number of especially useful Western techniques; but his three best candidates, have each failed to do so.[48] This failure occurred because efficiency and economic criteria dominated in devising "appropriate technology" and technology transfer which adapted people and cultures to technique, rather than the other way around.[48]

Once again Ellul recovered from this pessimism when computerization and new technology came into vogue. This gave him tantalizing possibilities for solving ACs' and particularly TWCs' problems. The *possibilities* he hoped for through these new techniques are decentralization, democratization, participation, liberation and transformation in the social production and distribution:[49]

The centralization of decision-making is necessary only to the extent that coordination and impetus demand it. In all other cases, the centralization of data processing can combine with a decentralization of decision-making. Decentralization is not only possible but facilitated: the computer relieves the decentralized collectivities of all-absorbing task and increases their decision making power by augmenting their means of information – data processing coordinates (and hence reinforces) the decentralized system, which, moreover, will soon be made necessary by the congestion of the center.

His hope is based on the synchronization of automation and computerization to permit vast network of automated, computerized and decentralized production units in which workers would have the time necessary for local management. Under this system workers are expected to control technique instead of being determined by it, their present unwieldy units being reduced to small, computer-linked groups of less than one hundred.[50] "Only if such changes are accompanied – not followed – by socio-political revolution affecting every structure in society the combined death-grip of a centralized bureaucratic state and the system being elaborated by technique be broken."[51]

Now Ellul's "possibilities" appear to be too remote and improbable. Evidence indicates greater centralization of existing institutional structures and mode of operations.[52]

SUMMARY AND CONCLUSION

We have begun this paper with the following proposition:

The world at present is undergoing various crises and among them a set of twin-crises stands out: a crisis of technology and a crisis of democracy. These two crises are linked and reinforce each other at a global level. The value premises behind these concomitant variable are not functionally and normatively synchronized.

This proposition has been examined at a global level on the basis of a dialectical paradigm consisting of productivist logic and logic of needs. The finding is that the former logic dominates and undermines the latter.

The underlying philosophy behind technological diffusion among the advanced countries is the Schumpeterian *"process of Creative Destruction"*, where we "all" advance at the expenses of the "few" and the destruction in the process of creation is not only necessary but inevitable.

Since the mid-1970s, technical-industrial modernization through new technology among the ACs has been taking place in a *mindless and chaotic manner*. This process of modernization has no conception of the relationship between destruction and construction, modernization and liberation – a relationship that today can no longer be considered simply in "productivist" terms. While the imperative of modernization has lost its perspective, it has also dramatically increased the relative significance of those people, institutions, values, and conditions of life that bear the cost of modernization among the ACs as well as TWCs.

Policy concerning technology, education, family life, economic growth neither checks the process of modernization nor guides it towards socially meaningful goals. Nowhere has democratization been accomplished to the point where the private character of decisions concerning the volume, kind, point in time, location of investment and transfer of technology has been effectively transformed in the manner of democratic control.

Unless these contradictions are fully recognized and a serious attempt is made to create a synthesis towards a new international order and development strategy, the double-crises-in technology and democracy – in the North-South relations are bound to increase. The North-South problems are also inextricably bound up with East-West conflict whereby humanity itself has become a hostage of industrial and military technology.

The principles of the new synthesis toward a *non-exploitative development path* should be based upon *participation, decentralization, demilitarization, respect for cultural alternatives, sustainability* and *justice*. In order to achieve *transformation*, we have to *transcend the present technology-transfer strategy*.

NOTES

[1] See chapter 2 and 3 respectively for "paradigms" and "dialectical paradigms" in R.T. Holt and J.E. Turner (eds), *The Methodology of Comparative Research*, The Free Press, New York, 1970.

[2] Source: F.R. Sagasti and A. Araoz, *Science and Technology for Development: Planning in the STPI Countries*, IDRC, Ottawa, 1979, p. 118.

[3] A. Etzioni, *A Comparative Analysis of Complex Organizations*, New York, 1961, pp. 23–40.

[4] R. Dahl and C.E. Lindblom, *Politics, Economics and Welfare: Planning and Political Economic Systems Resolved into Basic Social Processes*, New York, 1971.

[5] See generally P. Blumberg, *Industrial Democracy: The Sociology of Participation*, Contable & Co., London, 1968.

[6] See Rudolf Meidner, "Our Concept of the Third Way: Some Remarks on the Socio-political Tenets of the Swedish Labour Movement," *Economic and Industrial Democracy*, Vol. 1, 1980, pp. 343–369.

[7] J.A. Schumpeter, *Capitalism, Socialism and Democracy*, Harper & Row, New York, 1962, pp. 81–87.

[8] Claus Offe, *Contradictions of the Welfare State*, edited by John Keane, The MIT Press, 1984, pp. 207–219.

[9] C. Offe, at p. 192.

[10] S. Muthuchidambaram, *Microelectronics Technology: An Industrial Relations Perspective*, A Federal Task Force Report, 1982, pp. 1–4.

[11] C. Roveda and Ciborra, "Impact of Information Technology upon Organizational Structures," in *Microelectronics, Productivity and Employment*, OECD, Paris, 1981, p. 136.

[12] For more details regarding the developments mentioned in this paragraph see: J.K.

Galbraith, "Thinking Ahead: The Way up from Reagan economics," *Harvard Business Review*, July-August, 1982, pp. 6–12; G. Thompson, *The Conservatives' Economic Policy*, Groom Helm, London, 1986; Y.S. Brenner, *Capitalism, Competition and Economic Crisis: Structural Changes in Advanced Industrialized Countries*, Kapitan, Washington D.C., 1984; J.P. Gordus et al, *Plant Closings and Economic Dislocation*, Upjohn Institute, Kalamazoo, 1981; J.D. Roessner and R.M. Mason, *The Impact of Office Automation on Clerical Employment: 1985–2000*, Greenwood Press, Connecticut, 1985. For an excellent analysis of shifts in industrial relations due to these changes see Jack Barbash, "The New Industrial Relations", in IRRA *Proceedings of the 1986 Spring Meeting*, IRRA, Madison, Wisconsin, 1986, pp. 528–533.

[13] The Brandt Commission 1983, *Common Crisis North-South: Co-operation for World Recovery*, Pan Book Ltd., London, 1983, p. 37.

[14] Reuter News, *Globe and Mail*, January 6, 1967, B-16.

[15] On technological change and defence spending see the statement by S. Ramphal, Commonwealth Secretary-General, *The Hindu*, International Edition, December 10, 1983, p. 9; also refer Negative Economic Growth, *Globe and Mail*, December 10, 1986, B-8.

[16] Op. cit. note 13, at p. 43 and 44.

[17] D. Goulet, *The Uncertain Promise: Value Conflicts in Technology Transfer*, IDOC/ North America, New York, 1977, p. 17.

[18] "Science and Technology – Promises and Threats", in Paul Abrecht (ed), *Faith, Science and the Future*, World Council of Churches, Geneva, 1978, pp. 21–23.

[19] *Development Dialogue*, No. 1, 1980, pp. 55–67.

[20] See V. Rittberger, "Science and Technology as an International Order and Development Issue: An Overview", in V. Rittberger (ed) *Science and Technology in a Changing International Order, The United Nations Conference on Science and Technology for Development* (UNCCSTD), Westview Press, Boulder, Colorado, 1982, pp. 5–48.

[21] F. Stewart, *Technology and Underdevelopment*, 2nd ed, Macmillan, London, 1978, pp. 123–13. Also see F. Sagasti, *Technology, Planning, and Self-Reliant Development: A Latin American View*, Praeger, New York, 1979.

[22] op cit, note 13.

[23] "What Now? Another Development", *Development Dialogue*, No. 1/2, 1975, p. 93. Emphasis Added.

[24] V. Rittberger, op cit, note 20, at p. 19.

[25] *Ibid.*, at p. 20.

[26] For more details regarding the role of the UN and its Agencies, see notes, 13, 17, 18, 20 and 21 and also see S. Amin et al, *Dynamics of Global Crisis*, Monthly Review Press, New York, 1982; S. Amin, *Imperialism and Underdevelopment*, Monthly Review Press; United Nations Conference on Trade and Development, *The Development Dialogue in the 1980s – Continuing Paralysis or New Consensus*, UNCTAD's 20th Anniversary Symposium, UN, Geneva, 1984.

[27] C. Levinson, *International Trade Unionism*, Allen and Unwin Ltd., London, 1972; Chapter 3. Also see W. Morehouse, "Technological Autonomy and Disassociation in the International System: An Alternative Economic and Political Strategy for National Development"; D.W. Chu and W. Morehouse, "Third World Cooperation in Science and Technology for Development"; F.R. Sagasti, "Financing the Development of Science and Technology in the Third World"; all of which in V. Rittberger (ed) *Science*

and Technology in a Changing International Order, op cit, note 20. For the role and issues related to TNCs among the OECD countries see A. Morgan and R. Blanpain, The Industrial Relations and Employment Impacts of Multinational Enterprises: An Inquiry Into the Issues, OECD, Paris, 1977.

[28] W. Morehouse, cited in note 27, p. 59.

[29] C. Levinson, cited note 27, at p. 57.

[30] Cited note 27, at p. 95; also see C. Alvares, "Development Against People", Development Forum, July 1978.

[31] Frances Stewart, Technology and Underdevelopment, Macmillan, London, 1977, p. 277.

[32] U.N.C.T.D., 12 May 1981; also see Advisory Service on Transfer of Technology, UN, Geneva, 1986. Also see C. Edwards, The Fragmented World, Methuen, New York, 1985, 7.3.

[33] See F.R. Sagasti, "Financing the Development of S & T in the Third World", cited note 27.

[34] Statement by Mr. Janos Nyerges, UNCTAD, The Development Dialogue in the 1980s, cited note 26 at p. 39 and 40.

[35] See the chapters on Sweden, Denmark, and Norway in S. Muthuchidambaram, New Technology and Industrial Relations: Policy and Practices in Selected Countries, Research Report, University of Regina, Canada, 1986.

[36] See Chapter 8, "Economic Planning in the Two Other Orbits," in G. Myrdal, Beyond the Welfare State: Economic Planning and Its International Implications, Bantam Books, New York, 1967. Also see R. Meidner, op cit, note 6.

[37] V. Rittberger, cited note 20, at p. 37.

[38] W. Morehouse, cited note 27, p. 52 et seq.

[39] Ibid at p. 55. Emphasis added.

[40] L.T. Wells, Jr., Technology and Third World Multinationals, ILO, Geneva, 1982. This is an excellent study on this issue and this section of the paper is based on this source. For the other Working Papers of the ILO's Multinational Enterprises Programme (MULTI), 1979–1983, see Appendix, pp. 30–32.

[41] R. Bean, Comparative Industrial Relations: An Introduction to Cross-National Perspectives, St. Martin's Press, New York, 1985; p. 196.

[42] L.T. Wells, Jr., op cit, note 40, at p. 7.

[43] Ibid. at p. 13.

[44] See H.V. Perlmutter, "The Tortuous Evolution of the Multinational Corporation", Columbia Journal of World Business, 4, 1969.

[45] Cited note 40, at p. 20.

[46] Jacques Ellul, The Betrayal of the West, (trans), Matthew J. O'Connell, The Seabury Press, New York, 1978, p. 4.

[47] He has written 36 books and 400 articles on this subject. See Joyce M. Hanks, "A Way Out In A No-Exit Situation? Jacques Ellul on Technique and the Third World," in P.T. Durbin (ed), Research in Philosophy & Technology, Volume 7 – 1984, JAI Press Inc., Greenwich, Conn., at p. 283.

[48] J.M. Hanks, cited above, at p. 276.

[49] J. Ellul, The Technological System, (trans), J. Neugroschel, Continuum, New York, p. 94–95.

[50] Hanks, op. cit., at p. 278.

[51] *Ibid.*
[52] S. Muthuchidambaram, *Microelectronics Technology* . . . cited note 10, Chapter 1, and J.M. Hank, cited note 47, pp. 277–283.

FRIEDRICH RAPP

CULTURAL ALIENATION THROUGH TECHNOLOGY
TRANSFER?

1. I should like to deal with my topic in general terms within its broader historical and cultural perspective. This approach is justified because modern technology is an outgrowth of history, and because the type of technology used is a characteristic feature of any given civilization. Furthermore, a wide perspective opens the way for contrasting our present situation with similar, if not identical circumstances. In this way it should be possible to shed more light on the specific features we are being confronted with. After all Wilhelm Dilthey's (1833–1911) remark that what man is he can learn only from history, applies by implication also to our understanding of modern technology.

2. Technology transfer from one society to another is by no means a recent or unfamiliar phenomenon. On the contrary. There are good reasons to assume that from the very beginning of mankind some sort of cultural exchange is the rule. As Claude Lévi-Strauss put it, the only doom, the only fault that may hit a group of people is to prevent them from developing their potentials to the full by keeping them isolated.[1] But it is equally true that an all too strong (and imperfectly assimilated) foreign influence may result in the destruction of an otherwise flourishing and coherent style of culture and a specific type of identity of the individuals and the society concerned. Excessive and improper transfer may result in alienation. Clearly in the life of societies as well as in individual life there is a danger of running into new evils while seeking to avoid old ones.

3. Humans share a fixed biological nature, but they have only a potential, and hence an open, cultural nature or essence. Certain functional needs (breeding, relations of kinship, communication, division of labour, hierarchies, knowledge, meaning) must be fulfilled. How this is accomplished, however, is an open issue. For this reason a new-born child can develop into a full-fledged adult only by imitating (and thus tacitly adopting) the cultural patterns of its society and by accepting its education. There is a broad range for the specific type of cultural style

249

Edmund F. Byrne and Joseph C. Pitt (eds.), Technological Transformation: Contextual and Conceptual Implications, 249–257.
© *1989 Kluwer Academic Publishers.*

which a society develops or adopts. All the material and non-material elements of culture, from language, habits, and religion to tools and crafts, are open to unilateral transfer or bilateral exchange. The procedures by which this transfer is brought about range from peaceful exchange, trade and migration to missionizing, espionage, war and subjugation. The history of technology is abundant with processes involving the exchange of material and procedural innovations. There is even a stronger emphasis on technological exchange in prehistoric than in modern times, since in reconstructing prehistorical events one has to rely on archeological discoveries that are confined to material relics. In principle, then, there is nothing unusual about technology transfer.

4. At this juncture an intriguing normative problem arises. It is generally taken for granted that in the processes of exporting or importing material and non-material innovations the 'superior' culture will prevail, thus giving rise to the progress often read into the history of mankind. Yet by means of which criteria can one judge that a certain culture is higher than another one? An offhand answer is that the fact that a culture prevails teaches us that it is of a higher rank. When taken for granted without further qualification this amounts to the assertion that 'simply prevailing' equals 'superiority'. A similar situation obtains in the theory of biological evolution, where the notions of 'survival' and 'fitness' are taken to have the same meaning, since in the history of evolution 'being fit' is equivalent to 'being able to produce offspring.'[2] In the struggle for life, winning is the highest value and thus the highest rank is granted to the species or the culture that wins. Strictly speaking, this amounts to the descriptional tautology 'who wins wins', supplemented by the normative statement that the winner is to be considered as being of a higher rank than is the loser.

5. On closer inspection things turn out to be more complicated. Through the centuries the nimble nomadic tribes of the Asian steppes again and again attacked and conquered the neighbouring agricultural town-dwellers in China, India, Persia, and Eastern Europe. But after a few generations they were always absorbed by the more sophisticated and elaborated culture of the settled societies. The conquerors took on the culture of the nations which they had defeated.[3] Very much the same happened when, during the migration of nations, the Germanic tribes took over the Roman Empire. Within a rather short time they were

completely romanized. The lesson is that the winner in terms of military power may be the loser in terms of culture and vice versa. Hence different scales of ranking must be considered with respect to different fields of human activity. In speculative, philosophical terms this is to say that ideas, values and the realm of the mind may turn out to be superior to physical capacities and the realm of matter.

6. This finding is relevant when it comes to judgments about technology transfer. The common denominator of winning in the struggle for life, for military superiority and for technological achievement is that all of these processes proceed in terms of power. The power in question refers to the physical, biological side of human beings. This is were technology belongs since it consists in deliberately shaping the forces of nature so that they directly or indirectly enlarge, amplify or extend the range of our muscles and our senses (and in the case of the computer even of our mind). This is most evident in transportation and communication technology; but in one way or another it holds good for all technological processes or systems, since their output must ultimately be brought down within reach of our senses in order to be useful to us.

The argument put forward here is not meant to imply that the tool approach toward technology, as first pointed out by Ernst Kapp in 1877,[4] is exhaustive and the only appropriate one. To arrive at a more comprehensive account clearly also the symbolic, aesthetic, social, political and ecological dimensions of technology must be considered. My claim is simply that within the line of reasoning pursued here it is the tool approach which comes to the foreground. This approach is indeed the key to explanations of the superiority of modern technology.

7. When it comes to making a choice between two competing technological processes or systems the selection is made in terms of superior power or efficiency. Any technological system is designed to fulfill a certain function and, as a rule, it is evident which system serves this purpose best. Due to this mechanism of selection, the history of technology can, roughly speaking, be described as an accumulative process, leading to higher efficiency. This point was stressed by Henryk Skolimowski,[5] though he later renounced this approach. Examples are abundant: for long distance travelling the airplane is more efficient than the stage coach, and for communication the telephone is more efficient than smoke signals. Unfortunately these criteria also apply to military

technology. A submachine gun is more efficient than a bow and arrow and an atomic bomb is more destructive than dynamite. Since the latest state of technology is the result of an historical selection process in regard to superiority and power one need not wonder that this state is indeed superior to any other known type of technology.

At this juncture another qualification is necessary. The historical statement of fact recounted here does not imply that the historical course which technology took is to be exempted from criticism. Counterfactually one is free to imagine other lines of development than those actually obtained: public transportation systems instead of passenger-car traffic, Zeppelins instead of airplanes, etc. To consider alternatives that have been neglected may stimulate new types of technological problem-solving not sufficiently heeded hitherto. In a similar vein one can rightly criticize large-scale resource consumption, and polluting types of technology. It is our task to identify undesirable types of technology and to improve or replace them by better types that have less undesirable side-effects.

But one has to bear in mind that, generally speaking, doing so will not break the paradigm of selecting the most efficient solution. The change in question will 'only' refer to the criteria applied and the context considered. And it is just this change of perspective that makes all the difference! The general fact that in a given situation one must make a choice concerning the types of technology considered most desirable for the future on the basis of antecedent developments is part of our human condition and hence not subject to change.

8. Since it is an inherent characteristic of technology that it pertains above all to our body, to our physical existence, or to the material side of culture, the superiority of an innovation in terms of power and efficiency is self-evident. The biological features to which technology appeals are shared by all human beings, independent of their otherwise perhaps highly divergent historical and cultural heritage, their social and political structure, their ideological background, etc. When dealt with only in terms of maximum efficiency, modern technology tends to reduce the cultural features of man to mere biological issues and to dissolve the historically developed cultural heritage. Modern technology is by its very nature hostile to historical tradition and to cultural diversity, since these are factors that impede 'progressive' technological change and 'efficient' unification. Joseph A. Schumpeter's wording,

minted for a capitalist economy, also applies to the dynamics of technological change. It is indeed a "process of creative destruction,"[6] the
point being that the creativity in question is aimed at better fulfilling a
very specific (biological) function. How far this process has already gone
can easily be realized by comparing the everyday world of the cities of
different cultures, say, in medieval times and in present days. In our
times, housing, consumer goods, traffic and energy supply systems as
well as communication networks exhibit technologically determined,
unified traits, whereas in the Middle Ages the picture was dominated by
cultural divergencies.

9. In order to make this argument more convincing a further link is
needed, namely the close connection between technology on the one
hand and the non-material features of culture on the other. Why can
traditional features of culture not be maintained if a certain type of
technology is transferred and put to use? After all, the lesson of
premodern technology transfer is that certain technologies were taken
over without resulting in an overall, unifying effect. The answer is that
with modern technology things are different. Traditional technology was
embedded in the organic-biological sphere (riding animals and beasts of
burden, crafts, the strength of human muscles, the energy of the wind
and flowing water) and thus it could easily be adjusted to a human pace
and to the rhythms of life. Not so with the inorganic-mechanical processes induced in the artifacts of modern technology. They follow a logic of
their own. Modern technological processes and systems are designed in
such a way that we must conform to their functional principles (division
of labour, standardization, adjustment to recurrent mechanical processes, shiftwork, etc.) if we want to obtain their maximum output. We
created material devices to extend our physical capacities. The resulting
technological systems are detached from our body and have an existence
of their own. So we can no longer handle and control them in the same
immediate way as we control a hammer, for example, that acts as a
direct extension of our body. The price we have to pay for the efficiency
of modern technological systems is that we, in a way, allow them to
impose constraints upon us. Clearly the occupation of a coachman or a
blacksmith yielded a larger range for cultural variation than does the
work of a jet pilot or of a worker in an automobile factory. Modern
technology is so highly developed that its implementation demands
complete concentration and adaptation to the technological processes in

question. The unifying force is, as it were, built into the structure and processing of modern technological systems. This holds good not only for professional life but also for the sphere of consumption. Dealing daily with supermarkets, highways, airports, frozen foods and television will unavoidably lead to a specifically technology-shaped consumerist attitude towards life.

10. The line of reasoning followed here seems to support a materialist theory of history. In my argument it is taken for granted that technology is an integral part of the material life of a society and that modern technology will strongly influence the non-material features of a culture. Yet, my emphasis is more on the everyday experiences in the life-world than on the social, political and spiritual life in general. Everyday experiences pertain to rather tangible phenomena like standardization of taste, mass-media and bureaucratization. Indeed, they are largely determined by the very implementation of technological processes. With more abstract political and ideological issues there is a larger scope for variation, as can be seen, for example, by comparing Western countries with the Soviet Union and Japan. The differences obtaining here lend themselves to an explanation in terms of an idealist philosophy of history. The lesson is that one-dimensional, simplified schemes of historical explanation can hardly do justice to the complexity of the real world.

11. The conceptual and the material features of modern technology are the outcome of Occidental history. From a historical perspective the modern type of technology that tends to create a globally unified, technology-dominated culture of the life-world, is a recent phenomenon. It originated during the Industrial Revolution, which started about two hundred years ago in England. And since that time technological change has been going on at an ever increasing pace, so that it would be more appropriate to speak of a permanent revolution rather than of a finished process. Clearly *ex post facto* the potentialities for and the roots of this development can be traced back to much earlier times. The result is that modern technology accords with the Western world. Western culture is the natural environment into which modern technology was originally built. The idea of power over nature fostered by Francis Bacon and Descartes, the notion of progress fostered by the Enlightenment, as well as Max Weber's thesis that the Protestant ethic of striving

for professional success is the basis for a capitalist economy are at issue here.

12. Japan and Taiwan are among the rare examples of a 'successful' transfer of modern technology. But from the very tension within Japanese culture, i.e., the discordant amalgamation of modern technology transfer of modern technology. But from the very tension within Japanese culture, i.e., the discordant amalgamation of modern technology and traditional values, it can be learned that a real synthesis is by no means easily achieved within short terms. Inspired by the "Economic and Philosophic Manuscripts" of the young Marx, many Neomarxists take it for granted that in industrialized (capitalist) countries the technology-determined type of labour is a source of alienation from a self-determined, meaningful way of life. But compared to this type of, as it were, internal alienation in the industrialized nations, the forced transfer of modern technology into a cultural milieu that is shaped by a perfectly different historical heritage, and in which other values are cherished, must be considered to be a kind of collective alienation of a higher order of magnitude.

13. By no means does this imply that the Western countries are entitled to be forever the most 'successful' in implementing modern technology. Actually we can observe different forms of countercultures and alternative movements whose common merit is that they point in an exaggerated way to many faults and defects in our culture. If the hedonistic emphasis on pleasure instead of work and the plea for refusal instead of co-operation were to gain influence, it might turn out that in the future more disciplined and more ascetic nations will take over: Japan is an example in question. (The sceptic might object that this will presumably go on only until these nations in their turn become spoiled by wealth and luxury, in accordance with an historical law of rise and decline.)

14. Clearly in all of the problems mentioned here the pace of the change obtaining is a decisive variable. This is trivially true since in individual life as well as in historical processes a certain time is needed in order to adjust to a new situation and to integrate new elements into the existing patterns, thus bringing about a new synthesis. As for technology transfer the question is whether the all too rapid change forced onto developing countries by modern technology can be managed within a

foreseeable lapse of time, say within one or two generations.

15. Special problems demand special measures. In developing countries appropriate, soft, alternative, labour-intensive technology is needed and not the most sophisticated, capital-intensive high-tech. This does not imply that one has to go back to a former, outdated stage of technological development. Surely nobody would be willing to renounce the stock of the skill, knowledge and technological equipment attained at present, in order to bring about soft technology.

16. To state existing trends and to analyze functional relationships is one thing, but to evaluate these features is another. Thus, if one accepts the main line of the analysis of the facts put forward here, the question remains open whether one should welcome current trends and accept the unavoidable element of collective alienation implied in technology transfer. Or should one rather try to change the trends into some other direction? What is the appropriate normative judgment concerning the tendency towards a global, unified technology? Whoever seeks to answer this question must, at least implicitly, have an opinion about the desirable course of universal history. Clearly this question transcends human capacities. (It is not by chance that Whitehead called Hegel's attitude that of a god). This notwithstanding one can try to answer on a more modest, more practical, more provisional level. After all, whatever one does amounts to a choice. And if the choice is made deliberately it implies some conscious normative judgements.

Shall we then accept modern technology as the winner and classify cultures with a deviating heritage as losers? Clearly not! As pointed out in (5), material superiority does not by itself amount to overall cultural superiority. On the contrary. The quest for power and efficiency tends to absorb all efforts so that no resources are left for the unfolding and fulfillment of the symbolic, aesthetic, and intellectual dimensions of human existence. Putting too strong an emphasis on technology tends to disturb the ever intricate balance between an extrovert and an introvert attitude which is characteristic of a specific cultural pattern.

17. At this juncture we can observe a strange ('dialectic') feature of failure by success. Modern technology is designed to release man from the burden of hard work. Machines should take over in his stead, so that his freedom can be increased and he can turn to other, more rewarding,

types of activity. Now, however, modern technology is too successful in increasing efficiency and power. And it is this very intrinsic success which prevents the ideal of freedom from coming about. This for two reasons. First, the level of aspiration increases faster than technology can realize its fulfillment. As a result people tend to feel subjectively less satisfied in spite of the objectively higher standard of living available. Second, the benefit of our technology-shaped world is not free of charge. We have to pay for it by conforming to the structure and the operational principles of the technological processes and systems we have brought about and on which we rely. So there is the danger that when importing technology the developing countries may also buy these negative traits, in the long run even including some sort of failure by success. The lesson is that one should not refrain from importing technology, but that one should from the very beginning try to reduce the concomitant phenomena to a minimum.

18. Analogous considerations hold good with regard to the unifying traits of modern technology. We need different cultures. It is not sameness, but rather a diversity of cultural styles acting upon each other that makes up the richness of life on our planet. Creativity is fostered not by unification, but rather by exchange, mutual fertilization and competition. A diversity of cultures mutually influencing each other is the best antidote against a rigid, uniform, global technological civilization as well as against any type of technology-forced alienation. In this context the Eastern Buddhist attitude of non-doing and meditation could well act as a counterbalance against the Western, Promethean drive for activity.

NOTES

[1] *Race and History*, Paris (Unesco) 1961, Ch. 9.
[2] David Hull: *Philosophy of Biological Science*, Englewood Cliffs, N.J. (Prentice-Hall) 1974, p. 67.
[3] René Grousset: *The Empire of the Steppes*, Brunswick, N.J. (Rutgers UP) 1970.
[4] Cf. F. Rapp: *Analytical Philosophy of Techonology*, Dordrecht (Reidel) 1981, p. 4f.
[5] "The Structure of Thinking in Technology," *Technology and Culture* 7 (1966) pp. 371–383.
[6] *Capitalism, Socialism and Democracy*, New York (Harper) 1974, p. 81.

KRISTIN SHRADER-FRECHETTE

RISK AND TECHNOLOGY TRANSFER:
EQUAL PROTECTION ACROSS NATIONAL BORDERS

1. INTRODUCTION

Of the one million current and former U.S. asbestos workers between 300,000 and 400,000 have died, or are expected to die, of cancer. In the wake of these fatalities, the U.S. Occupational Safety and Health Administration (OSHA) has mandated workplace standards for asbestos, but rather than installing the cleaner technologies, many U.S. corporations, like Amatex, are continuing to use the dirtier technologies and moving their asbestos operations to other countries, notably Mexico. There are no Mexican regulations to protect workers from asbestos, dust levels in the Mexican plants are not monitored, and workers wear no respirators. Employees receive minimum wage, and are told nothing about the hazards they face.[1]

Asbestos processing is not the only technology transferred from the U.S. Millions of tons of U.S. hazardous wastes are now legally scattered throughout third-world countries that have been paid to take the toxics,[2] and massive advertising campaigns by corporations such as Dow and Chevron have turned the Third World into both a market and a dumping ground for dangerous pesticides, especially DDT. According to the U.S. General Accounting Office (GAO), 29 percent of all U.S. pesticide exports are products that are banned (20 percent) or not registered (9 percent) for use in the United States. The World Health Organization estimates that about 49,000 persons die annually from pesticide poisoning, many in the Third World, and that one person is poisoned by pesticides every minute in developing countries.[3]

The fundamental moral problem raised by each of these cases of technology transfer is whether corporations have an obligation to guarantee equal protection from risks, across national boundaries. Or, do they simply have an obligation to provide whatever protection is legally required in the country to which they export? In this essay, I sketch four main arguments used to justify transfers of hazardous technologies. I call them (1) *The Social-Progress Argument*; (2) *The Countervailing-Benefits Argument*; (3) *The Consent Argument*; and (4) *The*

259

Edmund F. Byrne and Joseph C. Pitt (eds.), Technological Transformation:
Contextual and Conceptual Implications, 259–275.
© *1989 Kluwer Academic Publishers.*

Reasonable-Possibility Argument. Next I show why all of these arguments, except the last, can be defused easily. I spend the remainder of the essay both arguing for consumer obligations to halt transfer of hazardous technology and sketching possible actions required by such obligations.

2. THE SOCIAL-PROGRESS ARGUMENT

Many utilitarian moral philosophers, especially act utilitarians, are opposed to recognizing rights to equal protection from risks, across national boundaries, because they admit of no such rights, whether within a nation or across nations.[4] Because of their stance, they would likely hold some variant of what I call the Social-Progress Argument. This argument is that, although no one wishes to see third-world peoples killed or injured by asbestos, hazardous wastes, or banned pesticides, nevertheless recognition of rights to equal protection would interfere with social progress in the countries that need it most. Act utilitarians, like Smart, typically believe that more human suffering is caused by following principles of equal treatment than by attempting to maximize the social and economic well being of a majority of people.[5]

Act utilitarians following the Social-Progress Argument might point out, for example, that worker fatalities during the building of the U.S. westward railroad reached a peak of approximately 3 per thousand per year.[6] This death rate is three orders of magnitude greater than the current, allegedly acceptable level of regulated risk in the U.S.[7] Nevertheless act utilitarians might view the death rate as a necessary evil, as the price paid to ensure an allegedly greater good, western expansion and consequent economic growth bringing an extraordinary level of prosperity to a majority of U.S. citizens.

The main problem with the Social-Progress Argument is its central presupposition that there is no in-principle obligation to recognize individual human rights, and that there are ethical grounds for sacrificing the welfare of significant numbers of human beings for the sake of the majority. This presupposition is problematic because act utilitarians are forced to admit that, on their view, not every individual would be protected from capricious or expedient denials of justice.[8] Even if nothing else was wrong with the Social-Progress Argument, this admission alone seems sufficient grounds for rejecting it.

The admission is devastating to the argument, but for reasons that I won't spell out in detail here. In brief, these reasons are that discrimination (e.g., providing unequal protection) is unjustified, unless it works to the advantage of everyone, including those discriminated against.[9] If it does not work to the advantage of everyone, then the discrimination is unjustified because it amounts to using some persons (victims of discrimination) only as *means* to the *ends* of other persons (beneficiaries of social progress), something is obviously wrong.[10]

3. THE COUNTERVAILING-BENEFITS ARGUMENT

But if failure to treat people equally possibly can be justified on the grounds that this failure helps everyone, including those treated unequally, then perhaps there is another argument for transfer of hazardous technologies. I call this second argument, the Countervailing-Benefits Argument. Whereas the Social-Progress Argument is that one ought to sacrifice the welfare of persons in the set A for the welfare of the persons in the larger set B, the Countervailing-Benefits Argument is that costs to persons in the set A are offset by benefits to the same persons in set A. This second argument amounts to the claim that, although it would normally be wrong to transfer technologies known to cause injury and death, recipients of the risky technology are better off than they would have been without it. At least they have jobs, for example, or are able to put food on their tables. The argument is that a bloody loaf of bread is sometimes better than no loaf at all. Hence, transferring hazardous technologies does not, *on balance*, constitute harm.

Perhaps the most questionable presupposition of the Countervailing-Benefits Argument is that any cost is allowable, provided that the benefits to the same persons are greater.[11] However, one could easily challenge this presupposition by arguing that some costs, e.g., certain rights violations, are *preventable evils* that ought never be allowed, even for countervailing benefits. Bentham, of course, rejected the notion of moral rights, but other utilitarians such as Mill challenged this rejection.[12]

If this interpretation of Mill is correct, then classical utilitarian doctrine is not a hunting license allowing infliction of wounds, so long as they are offset by greater pleasures.[13] Rather, one is not allowed, on

Mill's view, to threaten another's security, his very ability to experience happiness. Were one allowed to do so, then maximization of net benefits could be said to justify the worst sort of masochism, barbarism, and sadism.

Apart from whether Mill and other utilitarians recognize a right to security, there are a number of nonutilitarian grounds for believing that all persons share an implicit social contract guaranteeing equal, basic rights to security and that there are no countervailing benefits which might justify failure to reconize these rights[14]: *First*, all persons possess the two essential powers of moral personality: a capacity for an effective sense of justice and a capacity to form, amend, and pursue a conception of what is good.[15] *Second*, individuals and national societies are not self sufficient, but exist within a scheme of social cooperation and mutual interdependence.[16] *Third*, the comparison class is all humans, and all humans have the same capacity for a happy life.[17] *Fourth*, free, informed, rational people would agree to a social contract based on treating all humans equally.[18] *Fifth*, equal treatment of all persons provides the basic justification of all schemes involving justice, fairness, rights, and autonomy.[19] *Sixth*, all law presupposes a social contract guaranteeing equal rights; law itself embodies an ideal of the same treatment for persons similarly situated and an ideal of equal opportunity for all persons willing and eligible to compete for certain goods.[20] *Seventh*, and most important, without basic and equal rights to security, it would be impossible for anyone to enjoy any particular right (e.g., to property) which is legally guaranteed.[21] Moreover, although I shall not defend this point here, it seems reasonable to believe that there are duties to help provide security to those outside our national borders, especially if recognizing such a duty would involve only a minor sacrifice on our part. On Singer's scheme, for example, reasonable and benevolent people ought not forego a chance to do great good for others, in order to avoid a trifling sacrifice, both because they are able to do so and because the cost is not great.[22]

If there are grounds for recognizing either a right to security or a duty to protect others from threats to their security, then the Countervailing-Benefits Argument could be wrong. It could be wrong to try to justify violation of rights to security in exchange for the economic well-being associated with the transfer of hazardous technology. But how does one decide when the threats to security are grave enough to say that they constitute violation of a right or a duty? Henry Shue[23] has some suggestions in this regard. In the case of Mexican asbestos workers, for example, their security is threatened because (1) the technology does *physical damage* to their life, limb, and vitality, not just damage to their lifestyle; (2) it damages them in a *life-threatening* way; (3) the technology damages them in a way that is *irreversible*; (4) the technology does damage that is *avoidably undetectable* (because people in such a situation are likely to be poor, and hence unlikely to have proper medical advice and examination); (5) it does damage which is *avoidably unpredictable* (because workers lack the technical information about the risk, even though their employers may have it); and (6) the technology induces damage having a *high probability of occurrence.*

Even if transfer of hazardous technology were not questionable on the moral grounds that it jeopardizes individuals' rights to bodily security, however, it might still be questionable on factual or practical grounds: perhaps the transferred technology does not bring great or *countervailing benefits* to its recipients, contrary to the claims of the corporations associated with the technologies. For example, an executive of Velsicol Chemical Company, defending his company's sales of the pesticide Phosvel after it was banned in the United States, said: "We see nothing wrong with helping the hungry world eat."[24]

One problem with such an argument, however, is that between 50 and 70 percent of pesticides used in developing countries are applied to crops destined for export.[25] Moreover, third-world victims of pesticides do not appear to benefit even indirectly from the transfer of the dangerous technology, since foreign-exchange earnings are often not used to improve wages, housing, schools, and medical care for farm laborers. Foreign-exchange earnings benefit farm workers and pesticide users only if "trickle-down" economic procedures actually improve the overall welfare of workers most subjected to the hazards of a transferred technology.

In light of these considerations, the Countervailing-Benefits Argument is questionable on both moral and practical grounds. One practical problem is that many of the benefits alleged to go to third-world peoples in exchange for hazardous technologies might be overestimated. The moral problems are that the Countervailing Benefits Argument leads to undesirable consequences (e.g., justifying masochism) and ignores classical emphases on rights to security.

4. THE CONSENT ARGUMENT

In response, one could easily argue that, even if the alleged technologies threaten individual security, nevertheless the threats have been consented to by the recipients of the technology. According to proponents of the Consent Argument, unless one denies the autonomy of native peoples, and that of their representatives who make import decisions, then one is bound to allow all requested technology transfers.

Clearly the acceptability of the Consent Argument is a function of whether recipients of technology transfer[26] accept these technological risks in situations of informed consent. Consider, for example, whether a pesticide worker in a third-world country is likely to understand the probability and severity of the risk he takes and whether he gives free consent to it. It is well known that, as education and income rise, workers are far less likely to remain in risky occupations. This means that workers in high-risk jobs are, more likely than not, to be both financially strapped and poorly educated.[27]

Moreover, the situations in which third-world peoples would be most in need of jobs are precisely those which appear to preclude genuine *free* consent to accepting those jobs. This is because, where employment is most needed by local residents, there are likely few real alternatives to high-risk jobs. In Mexico, for example, the unemployment rate is 50 percent.[28] In such a situation, it cannot be assumed that a third-world worker actually gives informed consent to his risky occupation. This is because his job choice was probably not made in the context of ethically desirable "background conditions," such as the operation of a free market and the existence of alternative employment opportunities. But without these conditions, it is not clear that truly free employment choices actually take place. As John Rawls put it, "only against the background of a just basic structure and a just arrangement of economic and social institutions, can one say that the requisite just procedure [for occupational and other choices] exists."[29]

To determine whether background conditions, necessary for free informed choice by the pesticide worker, are likely to exist, recall that worldwide, about 800 million people, one-fifth of the world's population, are deprived of all income, goods, and hope. They live primarily in India, Bangladesh, Pakistan, Indonesia, sub-Saharan Africa, the Middle East, Latin America, and the Caribbean. Another one-fifth to two-fifths of the world's population, above the one-fifth that Robert

McNamara has called the "absolute poor," are chronically malnourished.[30] Given a situation involving disease, malnutrition, illiteracy, and squalor – not to mention few job alternatives and an economy likely not diversified – it is questionable whether, even with perfect information as to the risks involved, most workers could be said to *freely* consent to working with hazardous technology.[31] Moreover, such consent is not likely to be truly *informed*, since the same third-world conditions that militate against free consent also militate against education.

5. THE REASONABLE-POSSIBILITY ARGUMENT

If the analysis thus far is taken to be persuasive, then all three arguments enlisted to justify transfer of hazardous technologies face serious objections. Given the difficulties with these three arguments, the most promising way of ethically justifying transfer of hazardous technologies seems to me to be the Reasonable-Possibility Argument. This argument is based on the ethical maxim, "*ought* implies *can*"; if corporations *ought* to be required not to transfer banned technologies to third-world countries, then this requirement must be one that *can* be achieved, that is reasonably possible.

The main reason for believing that it might not be possible for a corporation to introduce safer technology on its own, in the absence of mechanisms to control the behavior of competing firms, is that such an action could financially destroy a corporation. To say that firms ought to employ the safer technology, when a government does not require them to do so, is to impose an unrealistic, unreasonable, heroic, and self-sacrificial burden on a corporation. And morality does not require heroism, only justice.[32]

6. A MORAL RESPONSE TO THE REASONABLE-POSSIBILITY ARGUMENT

Despite the truth of the claim that morality cannot rest on heroism, many philosophers are likely to argue that it is both reasonable and possible – not heroic – to cease transfer of hazardous technologies. Henry Shue, for example, has two basic objections to the claim that requiring corporations to cease transfer of hazardous technologies amounts to requiring heroism. He maintains: (1) that no institution has the right to inflict harm, even to hold down production costs; and (2) that underdeveloped countries cannot be expected to impose strict environmental and technological standards because they are competing with other countries for foreign investment. Hence, reasons Shue, if corporations cannot be expected to put themselves at a competitive disadvantage, relative to other corporations, then nations cannot be expected to put themselves at a competitive disadvantage, relative to other countries. In other words, he argues that, if *ought* implies *can*, then this dictum applies equally to corporations and to countries.[33]

The problems with Shue's objection (1) are both that it is difficult to say at what point inflicting a higher *probability* of harm constitutes infliction of harm and that, in fact, we do inflict harm in the form of increased probability of risk, in order to hold down production costs in the U.S. Our pollution-control regulations are specifically designed to trade a certain amount of safety for a specific amount of savings in production costs. The typical norm, adopted by the Environmental Protection Agency, a National Academy of Sciences panel, the Nuclear Regulatory Committee, and other government groups, is that safer technology is not required unless the risk to the public is greater than a one in a million increase in the average annual probability of fatality. Allowable worker risk is typically ten times greater than that for the public.[34]

In the case of many technologies, U.S. corporations are merely required to keep environmental hazards "as low as is reasonably achievable." What is "as low as is reasonably achievable" is fixed on the basis of a "favorable cost-benefit analysis." In the case of nuclear technology, for example, if it costs the licensee more than $ 1000 to avoid an additional person-rem of exposure to the public, then it is not required to do so; so long as this cost is under $1000 for each person-rem controlled, the licensee must aim at reducing maximum radiation exposure to the public to 0.0005 rem per person per year.[35] Hence, according to current law, there is no absolute prohibition against harm (where "harm" includes increased probability of risk), in part because such a prohibition would be impossible to achieve in a technological society.[36] But if so, then Shue's argument (1), as it stands, is incomplete.

Shue's objection (2) also makes a reasonable point, but it contains a flawed assumption. This assumption is that because countries compete with other countries for foreign technological-investment dollars, just as corporations compete with other corporations for profits, therefore countries have no more responsibility (than do corporations) to protect their citizens' health and safety by regulating technology. The assumption is flawed because it puts countries and corporations on the same level, so to speak, so far as protecting citizens is concerned. Putting them on the same level is problematic, at least because corporations have to concern themselves primarily with promoting *private* interests, i.e., maximizing shareholders' profits, whereas nations, arguably, have to promote the *public* welfare. Moreover, citizens, by virtue of their citizenship, share an implicit contract with their country; this social contract includes, for example, the provision that, in exchange for citizens' paying taxes and remaining loyal, the country performs many services such as protecting the health and welfare of its citizens. Except in the case of the employer-employee relationship, there is no comparably strong contract between a corporation and host-country citizens. Because there is not, it could easily be argued that the greater responsibility for protecting its citizens' health and welfare belongs to the country, and not to the corporation, that imports pesticides or agrees to manage toxic wastes.

Moreover, consider the consequences which would follow if one were to accept Shue's objection (2), that corporations have no more responsibility to force use of safe technology than do countries in which those technologies are located. If this were true, and if corporations did not willingly accept this responsibility, then corporations would be more likely to do as they wished in the face of a government that was alleged to have little responsibility to protect its people. In other words, Shue's objection (2) would likely lead to a self-fulfilling prophecy: less control of corporate behavior.

Apart from Shue's two objections, there might be other ethical reasons for questioning the Reasonable-Possibility Argument. Regardless of the precise nature of these objections, they all would likely come down to the same point, already mentioned in section 3. This is that corporate solvency is not to be given priority over recognition of basic human rights to security. The troubling ethical question raised by this point, however, is whether it is possible (without heroic sacrifice) for corporations to stop transfers of hazardous technologies.

7. A PRACTICAL RESPONSE TO THE REASONABLE-POSSIBILITY ARGUMENT

One cannot offer a realistic ethical solution to the problem of technology transfer, if the solution puts responsible corporations at the mercy of less scrupulous firms. Apart from what is ethically desirable, one cannot realistically expect corporations to do what governments, corporate employees, and consumers do not force them to do. But if not, what ethical obligations do we have, as consumers, to force transfer only of the safest technologies?

There are at least three reasons why we citizens in developed countries seem to have a moral obligation to help prevent use of hazardous technologies in underdeveloped countries. *First*, as Henry Shue notes, we have a "responsibility through ability."[37] If we have the ability to make a positive difference in such situations, we ought to do so.[38] As was already discussed earlier, our obligations to help vulnerable third-world persons arise from a number of considerations, most notably from the fact that we are interdependent and not self-sufficient and hence share an implicit social contract.[39] We are thus obligated to help other human beings because we are able to do so and because they are human beings.[40]

This first argument, quite obviously, is built on the claim that we are obligated to help others because we are able to do so and because they are humans. But how ought one to specify the limits on such an obligation? One could reasonably worry that such an obligation imposed too many restrictions, and one could exclaim, "Look, I have my own life to lead and my own children to raise, and I ought to be free of the obligation to help third-world peoples by promoting transfer of only the safest technologies."[41]

As Fishkin formulates the objection, one is morally required to prevent great harm when he is able to do so and when the costs and risks to him are minor. If one has only a modest number of occasions to help others, says Fishkin, then the application of the principle (above) to prevent great harm is not excessively burdensome and does not restrict one's freedom of action. This principle ("minimal altruism," as he calls it), however, has the cumulative effect of imposing great burdens and severely restricting one's free choices. The result, says Fishkin, is "breakdown," or "overload."

The objection is obviously correct in the sense that there is an upper bound to the cost that can be said to be required of persons striving to help those who need more physical security. What conditions might describe this upper bound? *First*, individuals clearly have a right to pursue their own commitments, apart from the sacrifices that appear to be demanded by impersonal global morality. Obviously, however, if one subscribes to the notion of a transnational social contract among all humans, then he ought not to forego a chance to do great good for others, in order to avoid a trifling sacrifice, as has already been mentioned.[43]

Second, individual sacrifices appear to become more burdensome and hence less of a moral imperative when they set us, either individually or as nations, at a disadvantage relative to others who have sacrificed less. *Third*, Henry Shue's distinction between the *scope* and *magnitude* of justice also provides some clues for an "upper bound" on our obligations to third-world peoples.[44] With respect to *scope*, everyone on the planet may have rights and duties grounded in global justice, because we may all be said to share a social contract within which these rights and duties may be specified. This does not mean, however, that the *magnitude* of the duties imposed on us all is the same. This is because there are a number of considerations which limit one's obligations to bring about social change. One such limiting condition, in addition to the two already mentioned, is that those with duties to help insure the security of other persons have no less rights to security than what is guaranteed by the minimum rights of those who are in need. This principle is obvious on the grounds of consistency.

A *fourth* limiting condition is that justice ought to be said to require only what some normal, nonheroic persons can be convinced to do. In other words, it ought to be possible to convince persons with healthy self-interest that they ought to act in accord with the authentic demands of justice; if at least some persons (having healthy self-interest) cannot be so convinced, then it is questionable whether the proposed standard of justice is legitimate. This is because one is only bound to do what is possible to do.

A *fifth* limit on the costs which can be demanded, in order to achieve the security necessary for equal treatment of all persons, is a function of the degree of coercion required to bring about recognition of duties to guarantee security. Gains in security bought at the price of either bloody revolution or totalitarian enforcement are highly questionable, primarily because of the cost in lives and in civil liberties. "Sometimes an unbloody half loaf is better than a bloody loaf."[45]

What all these limits reveal is that citizens in developed nations cannot morally reject the duty of helping to insure the safety of citizens in underdeveloped nations, purely on the grounds that the obligations imposed are too great and hence impossible to meet. The clear response to this objection is that, although one cannot be expected to help protect all peoples, one can (as Henry Shue puts it) protect "a few at a time until it becomes too heavy a burden."[46]

In addition to our "responsibility through ability" to help third-world victims of technology transfer, there is a second reason for this duty. As Shue also argues, we have a "responsibility through complicity," because we have accepted lower inflation and lower prices for foreign-produced goods, two benefits bought, at least in part, at the price of health hazards for peoples in developing countries.[47] The main point of this second argument is that, since citizens in developed countries benefit from the technological hazards transferred to underdeveloped nations, therefore the citizens in developed countries owe a debt of compensation or reparation. Judith Lichtenberg formulates a similar argument:

Suppose we consider a relationship, R, between a developed country, D, and an underdeveloped one, U. It may be that both D and U are better off with R than without it (though, of course, we make the artificial assumption here that the state to which we compare R is just the absence of R, with nothing replacing it). But suppose that by any reasonable standard, D benefits much more than U, not just in the sense that D ends up absolutely better off but also that it is improved more incrementally as well. This accords with the claim that economic relations between rich and poor countries widen the gap between them even if those relations bring absolute gains for all. So D is benefitted more by U's participation than U is by D's. Here the principle of unequal benefit applies to show that D owes something to U by way of compensation, for D owes its advantageous position in part to U's participation.[48]

In addition to our "responsibility through complicity," to help third-world victims of technology transfer, there is a third reason for this duty. We have a moral and a prudential obligation to help prevent use of hazardous technologies in third-world countries because many of the associated hazards affect us in first-world nations. For example, over 15 percent of the beans and 12 percent of the peppers imported from Mexico violate U.S. Food and Drug Administration (FDA.) pesticide-residue standards, and half of imported green coffee beans contain measurable levels of banned pesticides. The U.S. General

Accounting Office estimates that 14 percent of all U.S. meat is now contaminated with illegal residues. The pesticide residue problem has become so great that all beef imports from Mexico, Guatemala, and El Salvador have been halted. Moreover, government investigators found that half of all the imported food identified as pesticide-contaminated was marketed without any penalty to the producers and without any warning to the consumers.[49]

What all these examples illustrate is that it is virtually impossible to protect even U.S. citizens from the hazardous effects of technology transfers to third-world countries. Hence, just as planetary interdependence establishes a moral basis for our obligation to help those in underdeveloped nations, so also that same interdependence establishes a prudential basis for our obligation to help ourselves by helping members of third-world nations.

But what is the precise nature of our responsibility to help third-world victims of transfer of hazardous technology? What are we obliged, precisely, to do and what do we have the ability to do? There is no space here for a comprehensive answer, but it appears that we do have the ability to make it more costly for firms not to use safe technology than to use safe technology. As both the U.S. case of boycotting non-union lettuce and as the third-world case of boycotting Nestlé products revealed, American consumers, suitably organized, can send corporations a message through their pocketbooks.

First, citizens can boycott the products of firms known to use unsafe technology abroad. *Second*, U.S. citizens could demand U.S. export controls, e.g., arguing for a return at least to the Carter procedures and abandoning the more lax Reagan export controls. *Third*, citizens could demand that the U.S. acknowledge that all countries have duties to protect the security of their citizens or to recognize rights to security. *Fourth*, citizens could lobby for stopping all forms of U.S. assistance to all governments not recognizing their citizens' rights to security.[50]

Fifth, and more specifically, citizens could force abolition of the Overseas Private Investment Corporation (OPIC), an agency receiving Congressional (public) funds to distribute to American firms located abroad; much of the OPIC money has gone to American subsidiaries locating unsafe technology in third-world countries.[51] *Sixth*, citizens could also help third-world victims of transfer of hazardous technology by urging the U.S. to differentially favor governments that are more supportive of strong, independent unions.[52] A *seventh* practical strategy

for helping third-world victims of transfer of hazardous technologies
would be to urge the U.S. to require the U.S. Agency for International
Development (AID) to promote environmentally sustainable development projects.[53]

8. CONCLUSION

If the analyses in the preceding paragraphs are correct, then we have an
obligation to "make a difference," to make it difficult for nations and
corporations to subject unwitting third-world peoples to transfers of
hazardous technology. Our obligation is a duty whose recognition has
been in short supply in recent years, especially in the U.S. However, as
Thucydides reminded us, if we succumb to the belief that "someone
else" will carry our political responsibilites for us, then the common
cause will imperceptibly decay.[54] If the arguments in this essay are at
least partially correct, then we as citizens and consumers owe it to "the
common cause" to help prevent some of the abuses of technology
transfer.

NOTES

[1] D.R. Obey, 'Export of Hazardous Industries', *Congressional Record*, 124, Part 15, 95th
Congress, 2nd Session, June 29, 1978, pp. 19763–19764; hereafter cited as: Export.
[2] H. Shue, 'Exporting Hazards', in *Boundaries: National Autonomy and Its Limits* (edited
by P. Brown and H. Shue), Rowman and Littlefield, Totowa, New Jersey, 1981, p. 107;
hereafter cited as: EH and *Boundaries*.
[3] J.T. Mathews, *World Resources 1986*, Basic Books, New York, 1986, pp. 48–49. See also
R. Repetto, *Paying the Price: Pesticide Subsidies in Developing Countries*, Research
Report Number 2, December 1985, World Resources Institute, Washington D.C., 1985,
p. 3.
[4] For the distinction between act utilitarians and rule utilitarians; see D. Lyons, *Forms
and Limits of Utilitarianism*, Clarendon Press, Oxford, 1967; Lyons also argues for equiva-
lence between the two positions. See also M. Bayles, ed., *Contemporary Utilitarianism*,
Doubleday, New York, 1968.
[5] J. Smart, 'An Outline of a System of Utilitarian Ethics', in *Utilitarianism* (edited by J.
Smart and B. Williams), Cambridge University Press, Cambridge, 1973, p. 72; hereafter
cited as: OSUE. See also J. Mill, *Utilitarianism, Liberty, and Representative Government*,
Dutton, New York, 1910, pp. 58–59; hereafter cited as: *Utilitarianism*.
[6] C. Gersuny, *Work Hazards and Industrial Conflicts*, University Press of New England,
London, 1981, p. 20.

[7] See K. Shrader-Frechette, *Risk Analysis and Scientific Method*, Reidel, Boston, 1985, ch. 5; hereafter cited as *Risk*.

[8] Smart, OSUE.

[9] J. Rawls, *A Theory of Justice*, Harvard University Press, Cambridge, 1971; hereafter cited as: *Justice*. C. Fried, *Right and Wrong*, Harvard University Press, Cambridge, 1978. A. Donagan, *The Theory of Morality*, University of Chicago Press, 1977. See also S. Benn, 'Egalitarianism and the Equal Consideration of Interests', in *Equality*, Nomos IX, The Yearbook of the American Society for Political and Legal Philosophy (edited by J. Pennock and 75–76; and W. Frankena, *Ethics*, Prentic-Hall, 1963, pp. 41–42.

[10] W. Frankena, 'The Concept of Social Justice', in *Social Justice* (edited by R. Brandt), Prentice Hall, Englewood Cliffs, 1962, pp. 10, 14; hereafter cited as: *SJ*.

[11] Shue, EH, pp. 117 ff.

[12] One might read Mill as holding that utilitarian priciples require adherence to rules conferring rights, and that such rules exclude a case-by-case appeal to the general welfare. Mill explains that the primary object of moral rights is security, which he calls "the most vital of all interests," "the most indespensible of all necessaries, after physical nutrition," and "the very groundwork of our existence" (Mill, *Utilitarianism*, ch. 5). He affirms: "to have a right, then, is . . . to have something which society ought to defend me in the possession. . . ." of (Mill, *Utilitarianism*, ch. 5, par. 25).

[13] Shue, EH, p. 122.

[14] See C. Beitz, *Political Theory and International Relations*, Princeton University Press, Princeton, 1979, and 'Cosmopolitan Ideals and National Sentiment', *The Journal of Philosophy*, 80 (30), (October 1983), 591–600; hereafter cited as: *PTIR* and CINS. See also Rawls, *Justice*. Finally, see H. Shue, 'The Burdens of Justice', *The Journal of Philosophy*, 80 (30), (October 1983), 600-608; hereafter cited as: Burdens.

[15] J. Rawls, 'Kantian Constructivism in Moral Theory', *The Journal of Philosophy*, 77 (9), (September 1980), 515–572. See also Beitz, *CINS*, p. 595.

[16] Beitz, *PTIR*, pp. 129–136, 143–153; and J. Lichtenberg, 'National Boundaries and Moral Boundaries', in Brown and Shue, *Boundaries*; hereafter cited as: NBMB.

[17] W. Blackstone, 'On the Meaning and Justification of the Equality Principle', in *The Concept of Equality* (edited by W. Blackstone), Burgess, Minneapolis, 1969; hereafter cited as: *CE*.

[18] J. Rawls, 'Justice as Fairness', in *Philosophy of Law* (edited by J. Feinberg and H. Gross), Dickenson, Encino, California, 1975, p. 284; hereafter cited as: Fairness and *PL*. See also H. Shue, 'The Geography of Justice', *Ethics*, 92 (4), (July 1982), 714, 718.

[19] M. Beardsley, 'Equality and Obedience to Law', in *Law and Philosophy* (edited by S. Hook), New York University Press, New York, 1964, pp. 35–36. I. Berlin, 'Equality', in Blackstone, *CE*, p. 33. W. Frankena, 'Some Beliefs About Justice', in Feinberg and Gross, *PL*, pp. 250–251. M. Markovic, 'The Relationship Between Equality and Local Autonomy', in *Equality and Social Policy* (edited by J. Feinberg), University of Illinois Press, Urbana, 1978, p. 83. Rawls, Fairness, pp. 277, 280, 282. Finally, see G. Vlastos, 'Justice and Equality', in Brandt, *SJ*, pp. 50, 56.

[20] J. Pennock, 'Introduction', in *The Limits of the Law*, Nomos XV, The Yearbook of the American Society for Political and Legal Philosophy (edited by J. Pennock and J. Chapman), Lieber-Atherton, New York, 1974, pp. 2, 6.

[21] See H. Shue, *Basic Rights*, Princeton University Press, Princeton, 1980, esp. p. 24; herafter cited as: *BR*.

274 KRISTIN SHRADER-FRECHETTE

[22] P. Singer, 'Famine, Affluence, and Morality', in *Philosophy Now* (edited by K. Struhl and P. Struhl), Random House, New York, 1980, pp. 485–488; hereafter cited as: Famine.

[23] Shue, EH, pp. 119–123.

[24] Quoted by D. Weir and M. Schapiro, 'The Circle of Poison', in *Environment 85/86* (edited by J. Allen), Dushkin, Guilford, Connecticut, 1985, p. 119; hereafter cited as: CP and *Environment*.

[25] Weir and Schapiro, CP, p. 119.

[26] Such as Mexican asbestos workers, Central Americans living near a hazardous waste facility, and African officials.

[27] See M. Jones-Lee, *The Value of Life*, University of Chicago Press, Chicago, 1976, p. 39. See also E. Eckholm, 'Unhealthy Jobs', *Environment* 19 (6), (August/September 1977, pp. 33–34. Finally, see W. Viscusi, *Risk by Choice*, Harvard University Press, Cambridge, 1983, p. 46.

[28] Shue, EH, 129.

[29] Rawls, *Justice*, p. 87.

[30] E. Eckholm, 'Human Wants and Misused Lands', in Allen, *Environment*, p. 5.

[31] See A. Kuflick, 'Review of Henry Shue, *Basic Rights*', *Ethics*, 94 (2), (January 1984), 320, for an account of the difficulties of describing "coercion" and "free consent"; hereafter cited as: Review.

[32] See Shue, EH, pp. 130 ff. for a similar argument.

[33] Shue, EH, pp. 131–133.

[34] See K. Shrader-Frechette, *Risk*, chs. 4–5.

[35] 10 *Code of Federal Regulations* 20, U.S. Government Printing Office, Washington, D.C. 1978, p. 182; 10 *CFR* 50, Appendix I, p. 372; and Nuclear Regulatory Commission, *Issuances* 5, Book 2, U.S. Government Printing Office, June 30, 1977, pp. 928, 980.

[36] Shrader-Frechette, *Risk*, pp. 125–127.

[37] Shue, EH pp. 135 ff.

[38] Later I shall discuss how one might make a positive difference; for now let us consider the source of our obligation and possible limitations to it.

[39] See section 3. for reasons for espousing the nonutilitarian notion of an implicit transnational contract.

[40] Lichtenberg, NBMB, pp. 80 ff.

[41] T. Nagel, 'Ruthlessness in Public Life', in *Mortal Questions*, Cambridge University Press, New York, 1979, p. 84.

[42] J. Fishkin, *The Limits of Obligation*, Yale University Press, New Haven, 1982. See also D. Lyons, 'Review of Fishkin's *The Limits of Obligation*'', *Ethics* 94 (2), (January 1984), 328–329, and Kuflick, Review, pp. 321–322 for a formulation of this objection.

[43] See Singer, Famine, pp. 485–488.

[44] Shue, Burdens, pp. 602 ff.

[45] Shue, Burdens, p. 607.

[46] Shue, EH, p. 135.

[47] Shue, EH, p. 136.

[48] Lichtenberg, NBMB, p. 91.

[49] Weir and Schapiro, CP, p. 119.

[50] See Shue, *BR*, part 3.; see also Kuflick, Review, pp. 322–323

[51] OPIC has used taxpayers' money, for example, to help a U.S. corporation, Abex, build

a substandard asbestos plant in Madras, India, and to underwrite a U.S. corporation-owned substandard smelting complex in Africa.

[52] Obey, Export, pp. 19763, 19765. See also Shue, EH, pp. 137–138, 144.

[53] J. Seiberling and C. Schneider, 'How Congress Can Help Developing Countries Help Themselves', in *Journal '86*: Annual Report of the World Resources Institute, World Resourses Institute, Washington, C.C., 1986, pp. 57, 59.

[54] *The History of the Peloponnesian War*, bk. I, sec. 141.

JOSEPH AGASSI

TECHNOLOGY TRANSFER TO POOR NATIONS

The present essay belongs to the realm of global politics. It takes it for granted that the cleavage between poor nations and rich nations is not merely the problem of the poor nations but of the whole human race, since it threatens the very survival of mankind, and in many ways, and at the very least, it affects adversely the quality of life everywhere on earth. We are generally sufficiently aware of this fact so as to conclude that foreign aid is not the preference of the interest of the poor nation over the interest of the rich nation, but rather an act well within the national interest of the donor as well. This was epitomized by John F. Kennedy's edict: we can afford to offer foreign aid and we cannot afford not to. Also, Kennedy was aware of the difficulty of granting foreign aid to the poor nations on a permanent basis, like a rich philanthropist's regular aid to the poor as practiced well within all traditional societies; hence, foreign aid must aim at helping poor nations achieve self-sufficiency, i.e., learn to reach high levels of production so as to be free of the need for aid. This, of course, means the transfer of technology. Yet this idea, that the problem of the cleavage between the rich and the poor is not just the problem of the poor, because it is also the problem of the rich, this idea is still not fully endorsed by all concerned with the matter at hand. For, most past efforts in this respect are generally judged as failures, and the failure is usually – indeed, almost exclusively – laid at the door of the recipient nation or its government.

Foreign aid is a failure, to take the simplest and coarsest model or test-case, when a government of a rich nation donates a certain sum of money to the government of a poor nation, and most of the sum goes straight to a Swiss bank, to a personal bank account of the head of government of the poor nation. Even in this extreme and simple case, to say that the fault is one-sided is rather foolish. Obviously, the test-case is a failure, since the money was intended for a public purpose but has been converted to a private one, and one cannot reasonably deny that the failure is a failure of the poor nation to make a good use of money donated to it by the rich nation. And yet, since the problem is everybody's, it is also the failure of the rich nation. In other words, very

277

Edmund F. Byrne and Joseph C. Pitt (eds.), Technological Transformation: Contextual and Conceptual Implications, 277–283.
© *1989 Kluwer Academic Publishers.*

obviously, not only is the usurping of the money in the test-case a blow to the national interest of the poor; it is also a blow to the national interest of the rich. Why is it not perceived in this way?

Obvious as our conclusion is, it is very hard to maintain. After all, the foreign aid in our test-case was given in the form of money, rather than in the form of goods and services and advice, because of respect for the autonomy of the poor nation. No intervention, no strings attached, was the slogan. Let us agree, then, that the national interest of the poor nation is to keep the rich nations out of its internal affairs, even if as a result the government of the poor nation is less efficient and less just than the government of the rich nation. It follows that the disappearance of the donation into the bank account of the head of the government of the poor nation, though surely against the national interest of the poor nation, in a sense is a part of a process which must be judged in accord with the national interest of the poor nation. It is preferable to the poor nation to lose its donation funds to its head of government rather than to allow the government of the rich nation to interfere in the internal affairs of the poor nation and see the donation funds put to a good use. It is at least conceivable that this is so. Let us assume that it is. It follows that there is a conflict of interest between the rich nation and the poor nation, or at least that such a conflict is possible and more than merely possible. If the diminishing of the cleavage between rich nations and poor nations is a concern high enough on the global agenda, and if a rich nation is unwilling to reduce its quality of life drastically, and if all that the rich country can do to further its interest is to offer money with no strings attached in order to reduce the cleavage, and if the national interest of a poor nation is served only by a mode of conduct which, as a side effect, regrettably leads to the loss of the donated funds to the Swiss bank account of the head of government of that poor country, then, no doubt, we have on our hand a conflict of interest between the poor nation and the rich nation.

It is hard to believe this. And so we must find fault in the presuppositions spelled out in the previous long sentence. This can be done and should be done. Before we go into this, let us observe that the test-case presented above is very coarse and unrealistic. One reason rich nations have felt helpless to fight against corruption in the recipient nation is that this corruption is ever so often initiated well within the donor nation, by corrupt members of government agencies and by corrupt corporations and concerns and producers and transporters who all belong to the rich nation.

These things have come out only recently, and what is known is but the tip of the iceberg. But I prefer to leave these matters alone. My purpose in this essay is not to fight corruption, whether in the donor country or in the recipient country. My test-case involves corruption only because everyone in the rich countries concerned enough with the problem of the cleavage between poor and rich nations knows that foreign aid, aimed at closing or at least diminishing that gap, is a failure; and they customarily declare this the result of corruption, proposing tacitly that corruption is almost entirely the monopoly of the poor, and implying that against corruption even the best of good will is helpless, and concluding that foreign aid is in no way going to succeed, so that we better stop offering foreign aid and use the money thus saved for better purposes. I do not mean to save foreign aid – I do not know if foreign aid is at all the right means to overcome the cleavage between poor and rich nations and I also do not know if corruption is the worst obstacle to the task of overcoming this cleavage.

My purpose here is to show that this thinking, which is very common, contradicts the basic assumption that the problem of the cleavage between poor and rich nations is everybody's problem. Any other sort of thinking, which may be more realistic, will a fortiori lead to the same conclusions: whatever the rich do in order to better the lot of the poor, they do it selfishly, and if they fail in that, then they thereby show the limitation on their own ability to execute the task.

This may sound as if the poor have nothing to do with the failure, so that the rich may now try again while utterly ignoring the attitudes of the governments of the poor. Yet this, of course, is both unacceptable and impossible. Some philanthropic organizations, especially missionary ones, do act in total disregard for all local politics. Their action is, consequently, usually devoid of political import. They may save lives and thus win our admiration; they may help some individuals or even some villages become economically self-sufficient. But this is a far cry from helping a whole economy reach modern standards, which is the concern being discussed here.

When we observe a realistic test-case, such as the recent case of the Ethiopian famine, it is clear that the relief materials did not reach the starving masses as efficiently as expected and as reported in the Western mass media. It is also clear that the obstacles were diverse, and corruption at the top was not one of them, though ample corruption occurred at the lowest level of administration. It is also clear that the relief, being mainly grains and medications rather than money, could be distributed only by the government or, still better, by the donors with the government's cooperation, where the donors were international relief organizations rather than other governments. It is also a very significant fact that the donor organizations have learned not to yield to the Ethiopian government's pressures and thereby to procure more cooperation from them today then earlier. Since the emergency was larger at the beginning than later, clearly, the improvement was the result not of an emergency alone but, and chiefly, of the fact that the donors had representatives on the spot, who cared about immediate success, at least above a certain minimum, and who learned how to negotiate with the Ethiopian government.

This story, this meagre positive result which was purchased at a very high cost, teaches us a few things which, once learned, could make most of the literature on the subject more than a year old seem antiquated and useless.

The first conclusion is that, as the problem is global, it should not be left in the hands of donors and recipients: international organizations, in which all nations, donors, recipients, neighbors, should cooperate, should be in charge. This amounts to an international income tax, as envisaged by Paul N. Rosenstein-Rodan, as well as an international body designed to effect means which will redistribute the globe's wealth with a greater equity, as also envisaged by Rosenstein-Rodan. This, of course, raises the question of authority which can in principle be answered by endorsing the idea of a world government. But this idea would be rejected at once – less on the ground that it may be undesirable than on the ground that it may be ineffectual, will have no force to levy taxes and prevent wars, etc. The existence of certain international agencies, some belonging to the United Nations, others more of a philanthropic character, especially the International Red Cross, are possibly better starting points. Yet how can they be made to levy taxes and impose reforms which will lead to the redistribution of the globe's wealth?

The question is not new, only its dimension is. The foundation of the modern nation-state fundamentally included the same process. Ancient empires were the mere extensions of tribal societies, often by force, by the means of the sword. The modern nation state did not evolve out of such extension but out of the emergence of new nations and of popular endorsements of the national interest above and beyond the local interest. What made citizens of a given city or a given alliance of cities accept willingly the authority of the metropolis and declare it the nation's capital? No doubt it was partly the recognition that the yielding of some of the short-term interest would serve well the long term interest. But it is also a fact that the endorsement of nationalism served in part the short-term interest. It provided, perhaps most significantly, national identity and awareness. But these, however significant, could not work without some considerations of interest – long-term but also short-term.

There are a number of thinkers who work on awareness. A recent example is the work of Willy Brandt, which is extremely significant in that it presents international cooperation as both urgent and relatively inexpensive. Indeed, if it will raise the level of international trust even by a jot, it will save – by the reduction of armaments – more than it will cost. I will not, however, discuss raising public awareness of the need for the international cooperation necessary for the redistribution of the wealth of mankind all over the globe. Rather, let me observe, foreign aid is these days still an unhealthy matter between donor and recipient, partly unavoidably so, especially when recipient is a client state playing a significant role in the military system of the donor state. That the foreign aid system is justified by non-interventionism, is a fact: the donor gives the money to the recipient's government on the pretext that it is wrong to interfere in the recipient country's internal affairs. Yet, as Sir Arthur Lewis has argued long ago, this way donor governments support tottering rotten governments which would have been overthrown but for the support they gain from the donor's government, financial and other. And we do not need Sir Arthur Lewis's insight any more. We know that the moment President Ronald Reagan of the U.S.A. hesitated as to his support of President Ferdinand Marcos of the Philippines, Marcos's government fell, to the surprise of his most ardent opponents at home and abroad.

It is therefore quite reasonable for non-aligned nations – not nations which call themselves non-aligned – and individuals to begin to demand that foreign aid should go partly to an international foreign-aid body, preferably a United Nations special organization. This should be, at least partly, a proof of a bona fide foreign aid. The international body should not just offer money or relief materials but should negotiate aid with recipient governments as well as with the agencies and corporations involved.

Once such a body is established and funded, it will be able to examine the possibility of raising taxes from the international community, especially a tax related to fuel consumption. But I need not go further into detail. The matter was raised recently by Shimon Peres of Israel, who showed the need for massive aid to the Middle East and who proposed that a new Marshall Plan be effected immediately. His arguments concerning the need are unquestionable. Yet his plan is plainly unworkable and hence stands even less chances then were it workable. There is no doubt that one needs workable plans, plans with a fighting chance, before one can go to the international community and to the general public of the globe for support. The first step, then, must be the institution of an agency to work out and to implement such plans. A world relief organization which will receive all or some of the foreign aid and voluntary contributions which are in effect already may just become the lever for such a change. The main idea to observe is that we need much forethought and much cooperation even in order to transfer grains from overflowing granaries to a starving people. Once we realize this, then we will not wait for the next famine but start acting now.

Let us consider the proposal made here, to start a discussing at once, among those concerned with the problem, the possibility of erecting an international agency to which a percentage of foreign aid now given should be allotted, and which should become an international income tax agency in due course. This proposal is not inherently different from the proposal not to give the governments of poor countries monies but to negotiate with them the mode of aid. In both cases we are speaking of the use of existing means and the possible modifications of these means to make them usable for given purposes. These are instruments of social technology; and I claim that in social technology we use given tools in order to forge other tools, just as in physical technology. The general attitude here does not differentiate rich nations from poor ones in principle, but it does so in practice.

Social theory is split two ways. Traditional theory, with economics as its paradigm, ignores all differences between societies and sees in scientific and technological progress the only universal human factor and the only one that matters. Modern theory went to the opposite extreme, with social anthropology as the paradigm. The sad case of the disintegration of the tribal society of the Australian aborigines became a standard myth. The introduction of steel axes by totally uninterested merchants destroyed the social structure associated with the production of stone axes. Moral: do not modernize. Of course, societies with structures more firm than the Australian ones resist innovation for the sake of social stability. Hence the idea that better a corrupt government usurping foreign aid funds than the intervention of agents of foreign governments. The Ethiopian case a priori fits neither case. The Ethiopian government is an innovation. The famine was in part caused by innovations created by modern techniques introduced by the modern Ethiopian governments – the previous one which began the process of deforestation and the present one which, not noticing the role of women in agriculture, made reforms which destroyed subsistence agriculture and, introducing Soviet methods of marketing of agricultural products, destroyed the large scale agriculture. The need of agencies to negotiate relief is the need to use existing tools to create new ones.

And we must create new tools. It is in the interest of all nations to distribute the globe's wealth, and this means tighter international cooperation on all levels, especially what Segre calls micro-cooperation, and this means the use of existing social and political institutions for the creation of a new international network linked to national ones, and on a permanent basis. The formation of this network will offer incentives for the creation of the international income tax agency. The idea that the problem of cleavage between rich and poor nations is not the problem of the poor only but of the whole globe can thus be used as a tool, as a two-prong instrument to redistribute the globe's wealth more equitably than now.

BIBLIOGRAPHY

Agassi, Joseph. *Technology*, Reidel, Dordrecht, 1985.
Agassi, Joseph. "Closing the Credibility Gap", unpublished.
Brandt, Willy. *Arms and Hunger*, Pantheon, 1986.
Rosenstein-Rodan, Paul. *The New International Economic-Order*, Boston University, 1981.
Segre, Dan V. *The High Road and the Low*, Allen Lane, Penguin, London, 1974.

MARIO BUNGE

DEVELOPMENT AND THE ENVIRONMENT

Nearly all of us wish our own countries to develop, but we do not all want the same type of development, partly because we often overlook the fact that development is multidimensional. Likewise, all of us wish to continue utilizing and enjoying the environment, but few of us know what to do to protect it; and those who know how to protect it without sacrificing national development are even fewer.

The only point on which all well informed and rational beings seem to agree is that, unless certain drastic and world-wide steps are taken quickly, the industrial and military growth will finish off nature and, in the process, our own species. In other words, there seems to be agreement in that the unrestricted growth of industry and military might is incompatible with the preservation of the biosphere.

How are the two conflicting goals, development and the protection of the environment, to be rendered compatible with each other? How can we improve our welfare without putting the very existence of our species in jeopardy? To search for and find rational and practical answers this question we must begin by clarifying the key concepts involved in our problem, namely those of development and preservation of the environment. Once these concepts are elucidated we may hazard a strategy for a kind of development without unnecessary, hence avoidable, destruction of the biosphere.

1. HUMAN SOCIETY AS A SYSTEM

Every human society is a system composed of four subsystems: the biological, economic, political, and cultural ones (Bunge 1979). The biological subsystem is composed of people related by kinship bonds; the economic one, of the producers, distributers, and consumers of goods and services; the political one, of the citizens; and the cultural subsystem is composed of the producers, distributers and consumers of cultural goods and services.

In turn, every one of the three artificial subsystems of society – the

285

Edmund F. Byrne and Joseph C. Pitt (eds), Technological Transformation: Contextual and Conceptual Implications, 285–304.
© *1989 Kluwer Academic Publishers.*

economic, political and cultural ones – is composed of subsystems. The economic system is composed of agriculture and cattle breeding, mining and manufacture, as well as of trade and finances. The political system is composed of the government, the political associations, and the mass of citizens who, without participating actively in politics, are subject to obligations such as paying taxes and observing the laws of the land. Finally, the cultural system is composed of the communities of scientists, technologists, workers in the humanities, artists, teachers, as well as by all who those have access to the goods or services supplied by those communities. The individuals who do not participate in the activities characteristic of subsystem X are X-marginal. Typically, the inhabitants of the underdeveloped countries are largely economically, politically and culturally marginal.

Every one of the above mentioned systems is a system proper, not just an aggregate or collection, because its components are strongly bound together by biological or social ties. And the various subsystems that constitute a social system are strongly related to one another by such relations as those of exchange and power. These close relationships among persons and social groups give rise to global or emergent properties – such as social structure and political regime – that their components lack. And the fact that society is a system has the consequence that every reinforcement or weakening of one of its subsystems has repercussions on the others. The various subsystems of a society advance, remain stagnant, or decay jointly – though rarely in step. For the same reason none of the subsystems of a society is its permanent prime mover. Sometimes one of the subsystems advances, or decays, and the others follow it, though always with some time lag; at other times another subsystem takes the initiative.

For example, a prolonged drought and AIDS are disintegrating several African societies; the acceleration of industrial growth, together with the softening of Franco's dictatorship and the subsequent introduction of political democracy, made the current cultural renaissance of Spain possible; the French, Russian and Chinese revolutions caused profound economic and cultural transformations; and the current diffusion of the computer is causing revolutions in industry, commerce and culture.

In sum, human society is a very complex system wherein the advancement, stagnation or decline of one of the subsystems drags the others along. Ignorance of this fact has a high price. For example, the suppression of modern biology under Stalin contributed powerfully to the stagnation of Soviet agriculture, and the absence of democratic liberties protected an omnipotent and largely incompetent bureaucracy. Another example: the American industrial and international policies over the past few decades is causing the general decadence of the nation: it is no longer the first technological innovator, or the first exporter of manufactured goods, or even the first military power in the world, and European science is following closely at its heels. If we wish a society to prosper as a whole we must cause all of its subsystems to prosper at the same time.

2. FIVE TYPES OF DEVELOPMENT

Because every human society is a system composed of four subsystems – the biological, economic, political and cultural ones – the development of any society can be biological, economic, political, cultural, or global. And, since the prosperity of every subsystem depends critically upon that of the other subsystems, we must strive for a global development, i.e., the strengthening of all four subsystems at the same time – though not necessarily their growth.

Biological development is just the improvement in the living, working and leisure conditions, enabling us to improve our well-being and thereby lengthen the life expectancy. Another factor of biological development is the decrease in the birth rate, for it facilitates personal welfare, shrinks social problems, and alleviates the pressure on the environment. But this cause of so many social benefits is in turn an effect of a rise in the standard of living and in the cultural level.

Economic development involves the development (though not necessarily the growth) of the various industries of goods and services as a result of the technical innovations based on scientific discoveries. However, economic development is or ought to be a means, not a goal: it is a means to better enjoy life. Therefore it must be subjected to certain constraints, such as abstaining from polluting the environment or from exploiting people.

Political development is equal to an increase in popular participation in the design of policies the implementation of which may effect large sections of society. In other words, political development is but the introduction or improvement of participative democracy in all sectors, private or governmental, and at all levels, from the family and the firm to the private association – particularly the political party – and the state. A social group where indisputable authority and oppression reign is politically underdeveloped. True, one cannot eat freedom, but it is also certain that without it creativity and initiative cannot flourish. In the long run there is no important innovation in art, science, or technology – nor, of course, in social life or in government – without a modicum of freedom of expression, association, and participation in making decisions affecting the individual and his society.

Cultural development is better access to the cultural resources of society: an increase in the chances of participating, if only on a part time basis, in cultural activities or, at least, of enjoying the products of such activities. (At this point it may not be otiose to repeat that modern culture is not limited to the arts and the humanities, but embraces the natural and social sciences and technologies, from physics to history, and from engineering to management science.)

We have then four types of partial development: biological, economic, political, and cultural (Bunge 1980, 1987). By combining these four one obtains the *global development* of society. This is preferable to any partial development because the one-sided advancement of any social subsystem, even though it favors that of the others, may cause serious imbalances – such as a democracy of the hungry, or a prosperity of the uneducated or the oppressed.

Given the multidimensional nature of social development, or sociodevelopment, a single development indicator won't suffice. We need a list or vector with as many components as aspects or subsystems of society. In particular, the GNP of a society is but one component of such a vector; other components are the life expectancy, the percentage of the citizenry that participates freely in political activities, and the consumption of cultural goods and services (Bunge 1981).

Briefly, the type of development that improves the lot of most people is global development, for being the only one capable of satisfying all of the basic human needs and aspirations, and the only one capable of leading to a just and balanced society.

3. THE ENVIRONMENT

The environment of a population of organisms is that part of natural or social reality with which the members of that population interact. As a result of the differences and interactions among biopopulations, a large number of ecosystems or communities, composed of organisms belonging to different biospecies, have emerged in the course of evolution. Every ecosystem is a subsystem of the biosphere characterized by a habitat as well as by the biopopulations inhabiting it. A few species have access to the entire planet. Among them is ours. But, in contrast to other species, ours is the only one that is endangering the survival of all, including itself.

Human action alters the environment in an unavoidable manner. Every organism alters its environment by consuming some of its components and adding to it the waste products of its own metabolism. Man adds also the waste products of industry and commerce. Therefore it is not a question of preventing the alteration of the biosphere: the question is to protect it in order to prevent it from becoming uninhabitable. Evolution, yes; wanton destruction, no.

In other words, every organism is an open system, for it exchanges matter and energy with its environment. Besides, in the course of this exchange the organism builds up or maintains a high degree of order at the expense of an increase in the disorder or entropy of its environment. (Local entropy may decrease only at the price of an increase in global entropy.)

All organisms are entropic towards the environment and inwardly negentropic. But man is entropic or dissipative to the highest degree; and, at the same time, it is the only known animal capable of learning that he does not have to soil his own nest more than is strictly necessary. For example, we all know, or ought to know, that it is not unavoidable that we should go on destroying the biosphere as an effect of the uncontrolled exploitation of natural resources or of war. If we wished, we could turn, from clumsy exploiters, into wise managers of the ecosystems within our reach.

The ecosystems interfered with by human action differ from the natural ecosystems in that, in addition to following natural laws, they are subjected to rules dictated by human needs or wants. Natural ecosystems are self-regulated: they are in dynamical equilibrium. That is, they may respond to exogenous shocks, as long as these are not cosmic catastrophes, by repairing themselves adapting to the new circumstances. For example, in far away seas algae, the plankton, the fish and other organisms are in dynamical equilibrium. This balance is so robust that the system may, within limits, withstand climatic disturbances and even some human aggressions.

For example, the radioactive fallout produced by the collapse of the Chernobyl plant in 1986, and which reached the Mediterranean, was absorbed by the zooplankton, the faecal pellets of which sunk within a few days to about 200 m., thus cleaning up the surface of the sea (Fowler *et al.* 1987). On the other hand when the Peruvian fishermen overdid the catch of anchovies despite the warning of biologists, the unexpected arrival of the warm current "El Niño" during the Christmas of 1972 sufficed to kill a staggering number of fish in a population that had already been decimated by man. This provoked the so-called anchovy crisis of 1973, which practically put an end to the fishing industry in Peru for ten years. (By the way, the use of anchovies to manufacture fish meal for cattle is a typical example of the bad management of the biosphere.)

Consider an even simpler example. If you place goldfish or aquatic turtles in a fish tank, at the end of one week the system will stink to heaven unless it is cleaned. On the other hand, in a natural environment the waste products are utilized by plankton and unicellular organisms, that prevent the accumulation of waste. In short, the ecosystems exploited by man may lose their dynamic equilibrium as a result of contamination or overexploitation. To prevent this from happening their management must be designed on the basis of mathematical bioeconomics. (See e.g. Clark 1976.) Greed and the desire of domination, both myopic, are taking us to an ecological disaster on a planetary scale. However, such disaster can be averted with the help of science and technology provided the political will is formed.

4. ROMANTIC ECOLOGISM

The ecological movement born in the 1960s is of course a response to the massive and fulminating degradation of the biosphere caused by industrial over-production, the arms race, and the accumulation of non-recyclable waste, such as radioactive ashes and most plastics. It is also a response, though an indirect one, to the continuation of nuclear blasts and the threat of nearly instant destruction of the biosphere by a total nuclear war, which would cause not only radioactive fallout and fire storms, but also such an enormous quantity of ashes and dust, that the planet would be darkened and become so cold that all complex organisms would die.

The ecoactivists of all colors have been warning against such a degradation of the environment, and some of them have also alerted us against the plans for a war that would end up in a nuclear winter. Their alarm calls are based on irrefutable data and uncontroversial forecasts, such as the contamination of entire regions by the accumulation of industrial waste; the gradual warming up of the atmosphere by the "hothouse effect" caused by the massive emission of carbon dioxide by internal combustion engines and many industries; and, more recently, the discovery of the huge Antarctic hole (the size of the USA) in the ozone layer that protects us from ultraviolet radiation, and which seems to be due to a large extent to the fluorocarbons contained in some sprays of domestic and industrial use. Besides, the ecoactivists warn us that the unregulated exploitation of non renewable resources, such as oil and minerals, as well as of the renewable ones, such as forests and fisheries, is disinheriting our posterity. In other words, they are telling us that all the natural resources are limited, and that some human activities are destroying them irreversibly. See figure 1.

But the ecoactivists are divided with regard to both the concept of ecological balance and the means to be adopted to avert the final ecological disaster. In fact, they can be grouped into two classes, which shall be called the *romantics* or *green*, and the *scientifico-technical*, or *green-grey*. ("Grey" from the color of thinking matter.) The romantics believe that man is destroying the biosphere for having too large a brain: they are irrationalists. (See the fascinating novel *Galapagos* by Kurt Vonnegut Jr.) On the other hand, the scientifico-technical ecologists blame greed and improvidence, i.e., a bad use of a good brain. Moreover, we green-greys are confident that only a good use of the brain can solve the ecological crisis, which is only a part of a global crisis.

The romantic ecoactivists hold that industry disturbs the ecological balance, which they conceive of as being static or at most in a stationary state. But evolutionary biology teaches us that the ecosystems do not stay in equilibrium: that, although there are periods of stasis, once in a while there are also changes in species caused by genic mutations, natural selection, and other factors. Rather than staying in equilibrium, the natural ecosystems evolve. In sum, the greens do not know enough ecology.(See Küppers 1982.)

The result of human overpopulation, overproduction and pollution, as well as of the arms race, is not the disturbance of an imaginary

Fig. 1. Available renewable (R) and non-renewable (NR) natural resources.

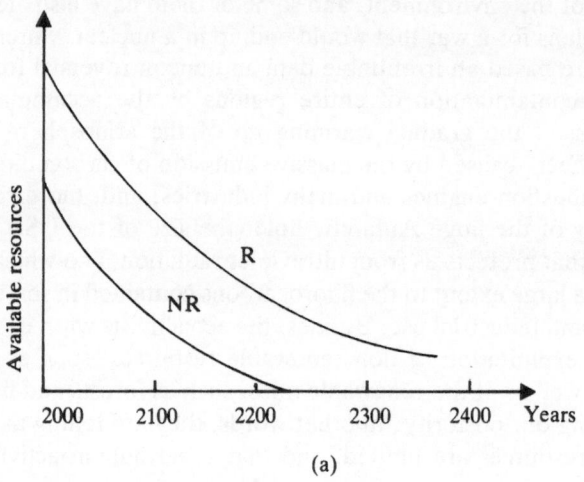

(a)

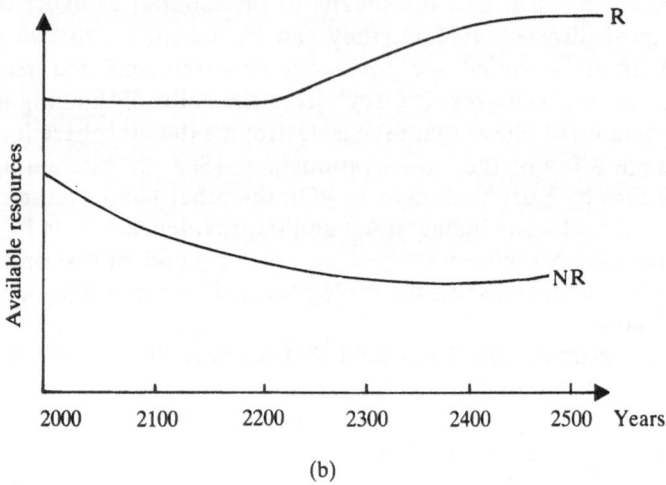

(b)

(a) Exhaustion of renewable (R) and non-renewable (NR) resources if they continue to be exploited at the present rate. Arbitrary scales. (b) Rational management of world resources: increase of renewable (R) resources up to the necessary level, and decrease of the rate of consumption of the non-renewable (NR) ones as an effect of restrictions on exploitation and substitution of fission energy for fossil fuels, and of wood and plastics for metals. Arbitrary scales ecological equilibrium but something far more serious. In fact, the result has been (*a*) the *destruction* of some components of the environment to the point of eliminating numerous biospecies, and (*b*) the *degradation* of the entire biosphere, in some cases perhaps irreversibly, endangering not only other biospecies but also and in the first place our own. *Homo sapiens sapiens* has become an endangered species in spite of having reproduced with an enormous success or rather, partly because it has multiplied out of all proportion with the natural resources. Like the broke nobleman, it is living beyond its means.

The massive and fulminous human aggression against the biosphere is over-ruling and sometimes replacing natural selection. Since artificial selection can be extremely quick and is usually in the service of myopic and selfish human interests, the natural controls and the natural sources of genic diversity are disappearing. For example, whereas the prehistoric peoples ate about 1,500 wild plant species, in the industrialized countries only about 80 species are cultivated for the market: the others are being neglected or even exterminated (Mooney 1979). In this way we are impoverishing nature and thus rendering it more vulnerable to climatic changes and pests. This is much more serious than a perturbation of the alleged ecological balance: we are hindering natural evolution and replacing it with an artificial evolution dictated by myopic and selfish goals.

The romantic or green ecoactivists propose a single heroic medicine to save ourselves from the ultimate ecocatastrophe: *Return to nature*, replacing modern industry with craftmanship, scientific technology with prescientific technique, science with the communion with nature, comfort with penury. In brief, they oppose development and preach underdevelopment. In this sense, the romantic ecologists are just as reactionary as Rousseau and some of the poets and philosophers of the Romantic period.

The great merit of the romantic ecoactivists is having opened our eyes to the race against nature that man started the moment he invented agriculture at the beginning of the Neolithic period, a race that has been accelerated immensely since the First Industrial Revolution. The great drawback of romantic ecologism is that it proposes doing the impossible, namely the return to a primitive stage in human history. History is an irreversible process. However, we can rectify the course: we can go onwards and upwards, instead of backwards and downwards. This is what the scientifico-technical ecologists, or green-greys, suggest we do.

5. SCIENTIFICO-TECHNICAL ECOLOGISM

Faced with the conflict between development and the defense of the environment, the romantic ecoactivists opt for the latter; that is, they propose sacrificing progress for the sake of the preservation of the biosphere. This proposal is neither reasonable nor practical for the following reasons.

Firstly, the return to a primitive stage in human history, were it possible, would cause a biological regression. In fact, without biology there would be neither medicine nor agronomy nor veterinary, without chemistry and biochemistry there would be no pharmacology, and without the latter we would fall prey to plagues of all kinds. The life expectancy could fall, in a single generation, to half the present one, i.e., the value it had one century ago.

Secondly, the languishing of science for lack of interest, of stimuli from technology, and of government support, would end up in a new Dark Age. No doubt, there would still be bards and composers, but there would be no scientific or technological researchers, and the technology and the industry necessary to diffuse the works of the bards and composers would be lacking. Life would become so empty and boring that people would not enjoy it, and many would take up drugs or prepare for death.

Thirdly, if modern culture, including technology, were to decay, so would the political system, to the point of tyranny or anarchy.

Fourthly, the goal of underdevelopment would hardly get the support of the masses, particularly in the nations of the Third and Fourth worlds. From the beginning of the Modern Age, and particularly from the start of our century, nearly all humans have legitimately aspired to maintain or improve their lifestyle. It would be impossible to persuade them that they must sacrifice this aspiration to save the peregrine falcon or the piranha. They would argue, and rightly so, that there must be an alternative to romantic ecologism and catastrophic development.

The rational and practical alternative is of course scientifico-technological ecologism. This policy may be summarized in one sentence: *Replace the uncontrolled exploitation of nature with its scientific management for the sake of all living beings except the parasites and the pathogenic germs.* In other words, instead of adopting the *laissez faire la nature* preached by the romantic ecoactivists we should design an interventionist though cautious ecological policy aiming at the preservation of the biosphere, ourselves included.

Whereas the romantic ecologist centers his or her attention on natural ecosystems supposedly in equilibrium (but in fact in evolution), the green-grey one centers his or her own attention on the systems accessible to human influence: these evolve not only by their internal dynamics but also in response to human work. This attitude is ecointerventionist, not ecoliberal, because its goal is to save the biosphere from all its enemies, human and non-human.

The intervention in the evolution of the ecosystems accessible to man ought to correct for the effects of natural catastrophes, such as droughts and floods. In addition, it ought to prevent (a) that certain biopopulations, e.g. of pathogenic germs, endanger the survival of others; (b) that the overpopulation of certain species, among them ours, ends up by killing them from lack of food; (c) that the industrial and military wastes continue to pollute the biosphere to the point of rendering it uninhabitable: (d) that industry exhausts with unnecessary haste the non-renewable resources, and deteriorates irreversibly the renewable ones; and (e) that development be accomplished at the expense of posterity.

The ecologists of both colors are agreed on points (c) and (d) above but not on the rest, since the green-greys propose that we intervene not only in society but also in nature. As a matter of fact this intervention is being carried out, though only on an exceedingly modest scale, in all civilized countries. For example, the foresters defend forests not only from fire and furtive loggers, but also from the forests themselves, by clearing them and hauling away the rotten logs. But they are powerless in front of acid rain, which is finishing off the North American and European forests and lakes; nor can they exert any control over the transnational firms that are quickly destroying the tropical forests of the world with the complicity of the local governments.

To put an end to ecological delinquency an ecological policy must be supplemented with a social policy. We must intervene not only in nature but also in society if we are to preserve the biosphere for our progeny. We must combine ecologism with global development, designing and implementing a policy of ecodevelopment and sociodevelopment on a world scale. (For the concept of ecodevelopment see Leff Ed. 1977.)

6. THE POLITICAL ASPECT OF ECOSOCIODEVELOPMENT

Since the biosphere covers the planet, if we wish to manage it to avert its further deterioration and restore it as far as possible, we must design and set up an *international agency endowed with effective authority to manage the biosphere*, or IABM for short.

Unless such an international management of the biosphere is set up above the national political sovereignties, we shall continue ruining it and thus ruining ourselves. Local measures, such as those taken by the canton of Basel to control the disposal of industrial waste, are necessary but insufficient when the problem is world wide – as ours is. For example, the talks on acid rain that the governments of the USA and Canada have been having for a number of years have yielded no results because the former claims to doubt the harmfulness of such rain. (It may be recalled that President Reagan stated, in the course of an election campaign, that certain "killer trees", not acid rain, are responsible for the massive destruction of the Northern forests. He did not care to mention the species to which those alleged murderers belong.) A powerful lobby of myopic vested interests is enough to cause the failure of any international talks destined to save "only" the biosphere.

Now, the creation of the IABM would be impossible unless it were äccompanied by deep and world wide social reforms. These reforms would have to be so many, and so radical, that they would amount to a *peaceful revolution* in all of the artificial social systems, i.e. the economy, the policy and the culture. I write 'peaceful revolution' because it would take at least one generation and because it could not be the result of violence. It is not a question of changing a handful of laws but a host of deep-seated habits. Nor is it advisable to resort to force, because force only generates resistance, and what we need to save ourselves is cooperation, not conflict.

Let us mention only a random sample of the system of social reforms needed to save mankind from self-destruction:

1. *Total nuclear disarmament* and dismantling all the laboratories and workshops aiming at forming new nuclear arsenals. Ensuring world peace by increasing the exchange of goods, services and people to the point when peace becomes more profitable than war.

2. *Gradual paralization of all the nuclear energy plants*, until we know how to dispose of nuclear waste and how to prevent that the plutonium generated in those plants be employed to manufacture nuclear bombs.

At the same time, vast investments in the research into alternative energy sources, particularly fusion. (A joint Soviet-North American-Anglo-German-French venture in this domain is bound to yield spectacular results before our millennium is over.)

3. *Final end to all aspirations to world domination*, whether political or economic, because it threatens world peace and cripples the development of the weaker nations.

4. *Strive for a more equitable distribution of the world's riches among all nations*, to decrease misery in the underdeveloped countries and minimize international conflicts. Reinforce genuine cooperation for development, the way Canada, Spain, Sweden and Switzerland have been doing. (See Spurgeon, 1979, and Pardos 1984.)

5. *Internationalize all the non-renewable resources and regulate strictly their exploitation*, taking posterity into account. In particular, tend to dispense with oil as a combustible, using it only to manufacture useful commodities. To this end, reduce gradually the production of oil and increase correlatively its price. In this way investments in the research into alternative energy sources, as well as in public transportation, will be stimulated.

6. *Rigorous regulation of industrial development* in order to (*a*) clean up the combustibles before burning them, (*b*) equip all industries with decontaminant and recycling devices; (*c*) prevent the waste of natural resources as well as of human ones (e.g. in the design and production of offensive weapons and junk); (*d*) abstain from producing goods and rendering services that are not strictly necessary to health and culture, aiming at null or even negative growth; (*e*) stress qualitative development: consume less goods and services but of better quality.

7. *Control the birth rate* – i.e. popularize family planning – to avoid infanticide, famines, and the existence of millions of homeless street urchins, decrease unemployment and the rate of exploitation of natural resources, and improve the quality of life of parents and children.

8. *Upgrade the educational system* on all levels, facilitating the access of the public to genuine culture, thus discouraging the massive consumption of junk in matters of food and drink, as well as of domestic and cultural products.

9. *Favor a more intensive public participation in the design of regional and international ecosociodevelopment*, so that everyone feels concerned about the preservation of the biosphere, the enhancement of the quality of life, and peace.

10. *Adapt technical innovations to the real needs of ecosociodevelopment*: (*a*) reject the purely cosmetic innovations, e.g. in the shape of containers, aiming at increasing unnecessary consumption; (*b*) develop not only high tech but also the intermediate technologies and even some of the crafts; (*c*) avoid both dwarfism (*Small is beautiful*) and giantism (*Big is beautiful*) in the design of industrial systems, public services, etc.; (*d*) decentralize or centralize on purely technological grounds and in the service of ecosociodeveloment rather than in obeisance to some ideology alien to it.

The pill we recommend is bitter, but not as bitter as the global disaster to which we are doomed unless we refuse to take it.

7. THE MORAL ASPECT OF ECOSOCIODEVELOPMENT

A revolution so profound, though peaceful, as the one we have just proposed, cannot be achieved overnight nor by decree. It will require at least one generation and plenty of lay saints like Ralph Nader and perhaps also religious ones like Mother Teresa – though such as help us live not die. It will also demand plenty of bold politicians, like Mikhail Gorbachev, ready to face inertia, pessimism, vice, and corruption.

Community and political action, though necessary, would be insufficient to achieve the results we want. We shall also have to introduce radical changes in the moral norms that rule our social behavior. The new morality we need must match the ideal of ecosociodevelopment, or global development together with respect for the enviroment and for our posterity. Once worked out, this ecosocial morality ought to be inculcated at the work place, at school and at home, particularly in the latter, which is where we learn early and from example.

Finally, the moral reform will have to be the basis of a law reform. For example, all constitutions ought to contain an article whereby the non-renewable resources are turned into international property, the management of which is put in the hands of the international resources management agency, which may farm out their exploitation to private enterprises or workers's cooperatives through temporary leasing contracts. Another example: the criminal codes will have to contain new concepts, such as those of ecological crime and embezzlement of items of international property. But of course the new laws, just like the old ones, ought to serve to deter and prevent rather than to punish. And, in order that they fulfill these functions, the legal sanctions ought to be accompanied by moral sanctions, without which there can be no respect for the law.

We have been writing about the new morality of ecosociodevelopment without specifying it. It is easy to say what it is *not*: it is not just a question of reinforcing or inventing moral rules guiding our behavior *vis à vis* the environment – e.g. "Thou shalt not litter" – but also norms concerning our behavior towards others, for the ecosociodevelopment revolution is basically a radical reform in our habits and aspirations and, therefore, of our conduct to neighbors and descendants.

We propose that the new morality should center on the principle *Enjoy life and help live*. However, in order to enjoy life and help others – in particular our offspring – we must have a place to live in: Hence we must observe the principle *Help preserve the environment*. In turn, this entails the end of the nuclear war threat – whence a further maxim must be observed as well: *Help nuclear peace*. Finally, the last two are necessary conditions for the survival of mankind, which everyone ought to be interested in adopting as the highest moral principle. See Figure 2.

Enjoying life is a right, and helping others enjoy it is a duty. In order that these two norms of personal morality may match with the ecosocial commandment concerning preservation of the biosphere, we should extend the right to life to all living beings except parasites and pathogenic germs. If we were to adopt this commandment literally, we should become vegetarians. (As a matter of fact most people do not eat meat because they cannot afford it.) This ideal is practicable because there are substitutes for animal proteins; besides, food scientists are likely to come up with artificial meat, and genetic engineering might be able to limit the production of bulls, roosters, and other males of the domestic animal species to the numbers strictly necessary for their preservation. The advantages of vegetarianism are : (*a*) a more efficient use of energy, and (*b*) a more consistent attitude towards other living beings.

Abstinence from meat is but one of the components of a long list of don'ts that we shall have to abide by if we wish to save the biosphere and improve our quality of life: avoid alcoholic beverages, tobacco, drugs that give momentary pleasure but harm health, and shopping for shopping's sake, as well as driving cars and producing and consuming junk. This proposal of frugal behavior should not be mistaken for an ascetic morality forbidding us to enjoy anything except doing business. Quite on the contrary, to enjoy life and help others enjoy it we need to practice

Fig. 2.   Sketch of a morality for ecosociodevelopment.

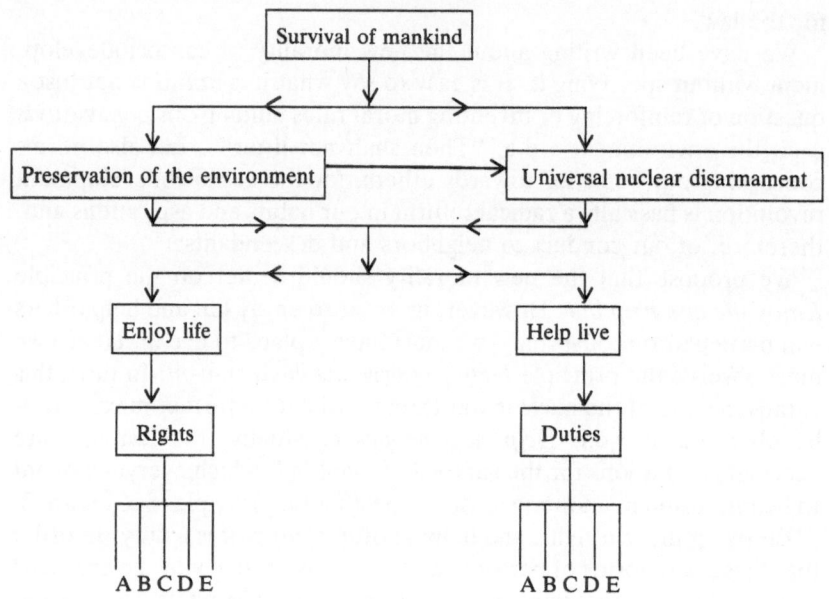

A = Environmental
B = Biological
C = Cultural
D = Democratic
E = Economic

continence in whatever may be harmful to ourselves or others, alive or
to be born. If we and our relatives and friends do not enjoy good health,
we shall not enjoy the outdoors or love, reading or the theatre, playing
or learning, working or engaging in community or political activity. The
question is not to stop pleasure but vice. And it is also a question of
reviving some moribund pleasures, such as those of the job well done
and the entertaining or instructive conversation.

The right to enjoy life, and the duty to help others enjoy it, can be
analyzed into five components each. These are the *environmental* rights
and duties, such as the right to swim in a clean lake and the obligation

not to pollute it; *biological*, e.g. the right to love and the duty not to contribute to overpopulation; *economic*, e.g. the right to work and the duty to do it conscientiously; *cultural*, such as the right to obtain information and the obligation to teach by example; and *political*, e.g. the right and duty of electing competent and responsible representatives and managers. Every right comes with a duty and conversely. And all the special rights ought to be subsidiary to the right to enjoy life, just as every particular obligation ought to be subsidiary to the duty of helping others live.

Every obligation limits some freedom. Only the absolute despot, or autocrat, can aspire to absolute freedom. The good citizen of a democracy knows that his freedom ends where that of his neighbor begins. Therefore economic neoliberalism – currently the partner of political neoconservatism – is the enemy of ecosociodevelopment and its attached morality, for preaching the boundless freedom to exploit human as well as natural resources. If we wish humankind to survive we must impose certain limitations upon economic freedom. But at the same time we must expand the cultural and political freedoms, as well as the means to make effective use of them. We must do this because only an informed citizenry capable of taking active part in the management of the *res publica* – which will increasingly become the *res mundi* – can protect the global and lasting interests and prevent the local and temporary from prevailing.

To sum up, in order to overcome the ecological, demographic and political crises, that is, to clean up and preserve the atmosphere, put an end to overpopulation and the resulting crowding, and eliminate the danger of nuclear omnicide, it will not suffice to adopt a mammoth set of economic and political reforms, both radical and world-wide. We shall also need a new morality. This morality cannot be the same as that of the romantic ecologists, who preach an impossible return to the state of nature or nearly so. (Man has never been fully natural: man is a tool maker and his own artifact, for he lives in a society that he makes and remakes all by himself.) The green-grey morality we have proposed is frugal but not ascetic: it recommends continence in some respects but also the enjoyment of pleasures that are becoming harder and harder to pursue in an increasingly filthy and tense world. The new commandments may be hard to follow in the beginning, but unless we learn to observe them not even the sinners will survive.

8. UTOPIA OR FUTURE?

Sancho Panza, the squire of Don Quixote, thought only of the here and now, of immediate advantage and enjoyment. He would undoubtedly object that what has been proposed above is utopian: that "people" won't accept it. No doubt, Sancho has a point: most people have not yet grasped how serious the ecological crisis is; they just do not know that, were the pollution and the plunder of the biosphere to continue at the present rate, the environment would become uninhabitable, in certain places at the end of this century, in others by the end of the next; and that a nuclear world war, because of the subsequent nuclear winter, would finish off all vestiges of life on the planet in a few days. In addition to having to face ignorance, negligence and irresponsibility, we must face the short-sighted vested interests of corporations that profit from an uncontrolled exploitation of natural resources that ought to be humankind's heirloom.

Precisely because of these motives, in particular ignorance and myopic interests, all responsible citizens ought to act quickly, demanding from their parties and governments that they study and implement drastic measures on a world scale. This is perfectly possible, as shown by the recent agreements on the elimination of chemical weapons, the limitations on nuclear missiles, and the production of sprays and plastics containing elements that are destroying the ozone layer. Further examples are the internationalization of Antarctica, the vigilance of the pollution levels in the Rhine and the Mediterranean, and the Soviet proposal, on the wake of the Chernobyl disaster, that the International Atomic Energy Agency inspect periodically all of the nuclear plants. This, then, is our reply to Sancho Panza: Although there is little time left to save ourselves, salvation is feasible, and as a matter of fact a handful of timid measures have already been taken.

However, even if Sancho concedes defeat, another enemy of ecosociodevelopment is ready to pounce on us: the smart skeptic, who will object that our proposal is utopian for ignoring the East-West rivalry as well as the class struggle. Our answer to this objection is as follows. Firstly, the ecological, demographic and nuclear crises are far more important than all the other conflicts for, without breathable air, drinkable water, and uncontaminated food in sufficient quantities for all, we won't survive; and without people nobody is going to participate in the East-West rivalry or the class struggle. Secondly, the ecological crisis is a result of the demographic, industrial and commercial explosions, which have shaken nearly all nations regardless of their political regimes. Lake Baikal and the Caspian sea are just as polluted as the Great Lakes and the Mediterranean, and the investments on antipollution devices are just as small in one bloc as in the other. Thirdly, the time is right because the world is no longer dominated economically, militarily or politically by a single world power or even by two. In fact, the world is becoming more and more multipolar, with the emergence of new centers of economic and political power, as well as of technological and scientific innovation, such as the European Economic Community, Japan, and several new industrial nations (Bellon & Niosi 1987). This depolarization of the world gives some chance to the initiatives that do not come from either of the two superpowers.

Briefly, the ecological and demographic crises are independent of the international and class struggles. Their solution is in the interest of everybody, regardless of economic status or political attitude, since the name of the game is staying alive, not gaining this or that advantage or winning this or that skirmish. The point is to cooperate on a world scale over and above the international and intranational rivalries. The above mentioned crises will be overcome to the extent that the required political will is formed. And for this to happen it is necessary to awaken those sho HavE not realized the seriousness and urgency of the problem; we must also build up, practice and preach an ecosocial morality.

To conclude, the alternative is not *either development or preservation of the environment*, but *either continuation of the course to ultimate disaster or improvement in the quality of life in and by ecosociodevelopment*. Unless we succeed in combining global development with the restoration and preservation of the biosphere, we shall deserve being rechristened, from *Homo sapiens sapiens*, into *Homo stultus stultus*. Worse, we shall deserve becoming extint for preferring profit and domination to the common and lasting good.

REFERENCES

Bellon, Bertrand & Jorge Niosi (1987) *L'industrie américaine fin de siècle*. Paris: Ed. du Seuil; Montréal: Boréal.
Bunge, Mario (1979) *A World of Systems*. Dordrecht & Boston: D. Reidel.
Bunge, Mario (1981) Development indicators. *Social Indicators Research* 9:369-385.
Bunge, Mario (1987) *Vistas y entrevistas*. Buenos Aires: Siglo Veinte.

Clark, Colin W. (1976) *Mathematical Bioeconomics: The Optimal Management of Renewable Resources*. New York: Wiley-Interscience.

Fowler, S. W., P. Buat-Menard, Y. Yokohama, S. Ballestra, E. Holm & V.Van Nguyen (1987) Rapid removal of Chernobyl fallout from Mediterranean surface waters by biological activity. *Nature* 329: 56–58.

Küppers, Bernd-Olaf (1982) Die Verlust aller Werte. *Natur* 4: 65–74.

Leff, Enrique, Ed. (1977) *Primer simposio sobre ecodesarrollo*. México: Associación Mexicana de Epistemología.

Mooney, P. R. (1979) *Seeds of the Earth*. Ottawa: Inter Pares.

Pardos, José Luis (1984) *Crecimiento y desarrollo en la década de ,os 80*. Madrid:Tecnos.

Spurgeon, David, Ed. (1979) *Give us the Tools: Science and Technology for Development*. Ottawa: International Development Research Centre.

BIOGRAPHICAL NOTES

Dr. Agassi is a Professor of Philosophy, Tel Aviv University and York University, Toronto (joint appointment). M.Sc. from Jerusalem; Ph.D from London School of Economics. His chief concern is to reform the commonwealth of learning so as to combat its current *exclusivity*. *Principal Works: Towards an Historiography of Science*, 1963, 1967: *The Continuing Revolution, A History of Physics From the Greeks to Einstein*, 1968; *Faraday as a Natural Philosopher*, 1971; *Science in Flux*, 1975; (with Yehuda Fried) *Paranoia: A Study in Diagnosis* ehp2.. 1976; *Towards a Rational Philosophical Anthropology*, 1977; *Science and Society: Studies in the Sociology of Science*, 1981; (with Yehuda Fried) *Psychiatry as Medicine*, 1983; *Technology: Philosophical and Social Aspects*, 1985; *Between Faith and Nationality: Towards an Israeli National Identity*, (Hebrew) 1983.

Albert Borgmann has been teaching philosophy at the University of Montana in Missoula since 1970. He is the author of *Technology and the Character of Contemporary Life*: (Chicago: University of Chicago Press, 1984).

Mario Bunge. Born in Buenos Aires in 1919, Ph.D. in theoretical physics, Professor of Physics, Universities of Buenos Aires and La Plata, of philosophy, McGill University, since 1966. Author of 31 books (70 counting their translations) and more than 300 papers on physics, philosophy of science, semantics, ontology, sociology, applied mathematics, etc.

Edmund F. Byrne, J.D., Ph.D., Professor of Philosophy at Indiana University – Indianapolis, concentrates on public policy issues regarding work and technology. His contribution to this volume, like "Building Community into Property" (*Journal of Business Ethics* 7, 1988, 171–83), foreshadows a forthcoming book, *Work and Justice: The Ethics of Automation and Globalization*.

Dr. Clifford Christians is a Research Professor of Communications at the University of Illinois-Urbana where he directs doctoral study in communications. Among his books, he is author (with Jay Van Hook) of *Jacques Ellul: Interpretive Essays* (Urbana: University of Illinois Press, 1981). He has been a Visiting Scholar in Philosophical Ethics at Princeton University and in Social Ethics at the University of Chicago.

John Crane is Professor of Social Work at the University of British Columbia. His recent publications include two monographs on research in social services and book on the evaluation of social policies. He is presently working on a book on applied social research, that draws on his 10 years experience in an advisory role to the government of Canada on welfare research projects.

Dr. Bernard den Ouden is Chairman and Professor at the University of Hartford. He is the author of *Language and Creativity*, 1975, *The Fusion of Naturalism and Humanism*, 1979, *Essays on Reason, Will, Creativity and Time: Studies in the Philosophy of Friedrich Nietzsche*, 1982. He has also edited *A Symposium on Ethics*, 1982, and *New Essays on Kant*, 1987. He has served as a consultant in planning and evaluation for various International Relief and Development organizations.

Paul T. Durbin, Professor, Philosophy Department and Center for Science and Culture, University of Delaware. Author of *A Dictionary of Concepts in the Philosophy of Science* (Greenwood Press, 1988). Editor: *The Reader's Adviser, 13th ed. vol. 5: Science Technology, and Medicine* (Bowker, 1988); *A Guide to the Culture of Science, Technology, and Medicine* (Free Press, 1980, 1984); 8 volumes in the series, *Research in Philosophy and Technology* (JAI Press, 1978-1985), and 3 volumes in the *Philosophy and Technology* series (Kluwer/Reidel, 1983, 1987, 1988). Member of the editorial board of Science, Technology, and Human Values and of the board of directors of the National Association for Science, Technology, and Society, among other boards.

James C. Klagge, Assistant Professor at Virginia Polytechnic Institute and State University, received his Ph.D. in Philosophy at University of California at Los Angeles. Klagge's research interests include moral philosophy and the nature of the good life, which motivated his contribution to this volume.

Tyrone Lai is Associate Professor in Philosophy at Memorial University, Newfoundland, Canada. His articles have appeared in *Historia Mathematica*, *Journal of the History of Philosophy*, *British Journal for the Philosophy of Science*.

Joseph Margolis is a Professor of Philosophy at Temple University and has just completed a three-volume overview of current problems in Western philosophy. The titles include: *Pragmatism without Foundations* (1986), *Science without Unity* (1987), *Texts without Referents* (1988) – all with Basil Blackwell.

Professor S. Muthuchidambaram is a Professor of Administration at the University of Regina. He is the author of *Microelectronics Technology: An Industrial Relations Perspective*. He concentrates on labor law and industrial relations, administrative law, and business ethics.

Joseph C. Pitt is Professor of Philosophy and Director of the Humanities, Science and Technology Program at Virginia Polytechnic Institute and State University. His main interests are in the history and philosophy of science and technology. The author of *Pictures, Images, and Conceptual Change*, numerous articles, and the editor of several volumes in history and philosophy of science, he is currently working on a book length manuscript, *How to Think About Technology*.

Friedrich Rapp is a Professor of Philosophy at the University of Dortmund (West Germany). Books: *Analytical Philosophy of Technology* (1980), editor of: *Contributions to a Philosophy of Technology* (1974), and together with Paul Durbin co-editor of: *Philosophy and Technology* (1983).

Kristin Shrader-Frechette, currently Graduate Research Professor of Philosophy at the University of South Florida, did her undergraduate work in mathematics and physics and received her Ph.D. in philosophy of science from the University of Notre Dame. Editor-in-Chief of the Oxford University Press series of *Monographs on Environmental Ethics and Science Policy*, Shrader-Frechette has published approximately 70 articles and 5 books/monographs: *Nuclear Power and Public Policy* (1980), *Environmental Ethics* (1981), *Four Methodological Assumptions in Cost-Benefit Analysis* (1983), *Science Policy, Ethics, and Economic Methodology* (1984), and *Risk Analysis and Scientific Method* (1985). She and her husband Maurice, a mathematician/computer scientist, have two children.

Lan Xue is currently working on his Ph.D. degree in the Department of Engineering and Public Policy at Carnegie Mellon University. Trained as a mechanical engineer, he received two master degrees in Technological System Management and Public Policy Analysis from SUNY, Stony Brook. He is a member of Society of Chinese Young Economist and author of a computer training textbook, *Jump up in DOS*.

TOPICAL INDEX

PHILOSOPHY AND TECHNOLOGY

Series Editor: Paul T. Durbin

OFFICIAL PUBLICATIONS OF
THE SOCIETY FOR PHILOSOPHY AND TECHNOLOGY

---

1. *Philosophy and Technology*
   Edited by Paul T. Durbin and Friedrich Rapp.
   (Published as Volume 80 in 'Boston Studies in the Philosophy of Science')
   1983, xiv + 344pp.   ISBN 90-277-1576-9
2. *Philosophy and Technology, II.* Information Technology and Computors
   in Theory and Practice.
   Edited by Carl Mitcham and Alois Huning.
   (Published as Volume 90 in 'Boston Studies in the Philosophy of Science')
   1986, xxii + 352pp.   ISBN 90-277-1975-6
3. *Technology and Responsibility*
   Edited by Paul T. Durbin.
   1987, x + 392pp.   ISBN 90-277-2415-6
4. *Technology and Contemporary Life*
   Edited by Paul T. Durbin.
   1988, viii + 320pp.   ISBN 90-277-2570-5
5. *Technological Transformation.* Contextual and Conceptual Implications
   Edited by Edmund F. Byrne and Joseph C. Pitt.
   1989, xii + 314pp.   ISBN 90-277-2826-7
6. *Philosophy of Technology.* Practical, Historical and other Dimensions
   Edited by Paul T. Durbin.
   1989, in preparation.   ISBN 0-7923-0139-0

Kluwer Academic Publishers
DORDRECHT / BOSTON / LONDON